AN ADVENTURER'S GUIDE TO NUMBER THEORY

AN ADVENTURER'S GUIDE TO NUMBER THEORY

by RICHARD FRIEDBERG

DOVER PUBLICATIONS, INC.
New York

To Leslie and Cynthia

Copyright

Published in Canada by General Publishing Company, Ltd., 30 Lesmill Road, Don Mills, Toronto, Ontario.
Published in the United Kingdom by Constable and Company, Ltd., 3 The Lanchesters, 162–164 Fulham Palace Road, London W6 9ER.

Bibliographical Note

This Dover edition, first published in 1994 is an unabridged and corrected republication of the work first published by The McGraw-Hill Book Company, New York, in 1968. In addition to corrections in the text, two new appendixes have been added to this edition.

Library of Congress Cataloging-in-Publication Data

Friedberg, Richard, 1935–
An adventurer's guide to number theory / by Richard Friedberg. — Dover ed., corr. & expanded.
p. cm.
Includes bibliographical references and index.
ISBN 0-486-28133-7
1. Number theory. I. Title.
QA241.F75 1994
512′.7—dc20 94-33113
CIP

Manufactured in the United States of America
Dover Publications, Inc., 31 East 2nd Street, Mineola, N.Y. 11501

CONTENTS

AN ADVENTURER'S GUIDE TO NUMBER THEORY

A	B	Γ	Δ
1	2	3	4
E	F	Z	H
5	6	7	8
Θ	I	K	Λ
9	10	20	30
M	N	Ξ	O
40	50	60	70

Π	ϙ	P	Σ
80	90	100	200
T	Y	Φ	X
300	400	500	600
Ψ	Ω	ϡ	/A
700	800	900	1,000
KF	ΣF	ΣΞ	ΞB
26	206	260	62

The Greek system of numbers relied mainly on letters of the alphabet. It required twenty-seven symbols and was still much more cumbersome than our system of Arabic numerals.

1. Seven jogged my elbow

Oil, Cake, and Glass

Every number has its character. I picture 7 as dark and full of liquid, like oil when it oozes from the ground. Three is lumpy and hard, but dark also, and 4 is soft and doughy and pale. Five is pale but round like a ball, and 6 is like 4 only richer, like cake instead of dough.

These are only my ideas. Perhaps you see 7 as a pincushion, and 5 as a bright spot of light. Everyone has his own pictures. Here are some more of mine.

Two is solid and tingly, like the Liberty Bell. Twelve is like a more delicate bell, of glass instead of iron. Eight is rough and hard like a stone, and 10 is smooth like a pebble on the beach. Nine is tingly but much sharper than 2 or 12, and it seems ready not only to ring but to shatter and burst like a fruit. The number 1 is not very interesting, about like 7 partly dried up, but when it is written "one" then it seems big and mysterious, ready to swallow everything up, the grandfather of numbers.

Perhaps you are surprised to read about numbers this way. They are usually treated much more dryly, because most mathematicians have forgotten about these pictures by the time they grow up. Writers might remember, but they are usually more interested in people, and think of numbers as impersonal and cold.

The larger numbers, to my mind, follow certain patterns. Usually odd numbers are darker than even numbers. Some odd numbers are wet, but all even numbers are dry. A two-digit number takes after its digits, but it also resembles its factors, the numbers that go into it evenly. For example, 92 seems like great heavy, smooth blocks. Now I think this is because 92 is a multiple of 4, which makes it squarish instead of round. But instead of being doughy, 92 is smooth and solid like 2, and heavy because of the 9. But 92 also makes me think of a movie called "The House on 92nd Street," and I used to live near there in Manhattan and seem to remember a house on 92nd Street made of heavy stone blocks.

All this depends on the cast of my mind. I rarely associate the numbers with colors. I incline more to textures—wet, dry, smooth, rough, heavy, light. When I was a child, I had no gift for art, but was fascinated by the way things felt to the touch. I used to run my hand over surfaces and learn the cool, restful feel of wood, the cold bite of metal, and the chill of glass.

Although I don't use colors, I do use light and dark. All numbers in the 20's seem flooded with sunlight. Numbers in the 30's are bumpy, in the 60's are very dry, and in the 70's are dark and mysterious. Numbers in the 40's don't seem at all doughy, like 4. It is the numbers *ending* in 4 that seem doughy, and the doughiest of all is 64, which is $4 \times 4 \times 4$. The doughiness of 64 has to do with its having no odd factors. You can divide it by 2, and divide it again and again until you get all the way down to 1 and have nothing left to divide by, as if you rubbed a piece of dough between your hands until it had all blown away. Sixty-six is also dry and powdery, but it is gritty, for you know that if you divided it by 2 you would get stuck at dirty 33.

Some of the most beautiful numbers are multiples of 5. Twenty-five is thin and brilliant and round, but I prefer 75 with its dark, rich glow, like a deep ruby or a moonlit evening. The roundness of both numbers comes from 5×5, and the light or dark from 2 or 7. Specially intense effects are produced by a multiple of some number which also contains that number as a digit, as 36 contains 6 or 64 contains 4. For example, 16 is rich because of the 6 and strong like the roots of a tree because it is 4×4. But $96 = 16 \times 6$ is rich and strong and also has an extra coloring from the 9. Seventy-two $= 12 \times 6$ has the bell-like quality of 12, but much deeper. One hundred and twenty-five is like 25, and 375 is like 75. The powers of 10 are like chiefs of different degrees. One hundred is a captain on horseback, and 1000 is a king, crusted with majesty.

Obviously, the multiples of 5 and 10 have special qualities only because we count by fives and tens. If you have studied the New Math, you know that we can just as well count by some other number, say by eights. Then the number $4 \times 4 \times 4$, which I think of as all made of dough, would be written 100 and would be sitting on horseback as a chieftain!

In any case, the relation of a number to its factors is the same, no matter how we count. To my mind, multiples of 5 are round, of 2 are dry, of 4 are strong, of 7 are sad, of 10 are bland, of 3 are juicy. But some numbers, even large ones, have no factors—except themselves, of course, and 1. These are called *prime* numbers, because everything they are starts with themselves. They are original, gnarled, unpredictable, the freaks of the number world.

Pythagoras and Ramanujan

All this is just the opposite of arithmetic, which treats all numbers alike and reduces them to a system. People say that when you are in the army you are treated as if you were a number. No one cares who you are or what you are like, as long as you appear in the right place at the right time. But the reason we call that "being treated as a number" is that we have all learned, in arithmetic, to treat numbers as soldiers. Arithmetic is the gospel of those who are interested in hard, cold facts, who don't care whether 7 is a prime number, but who know that 7 dollars are better than 6, and 8 better than 7.

Consider the fascinating numbers 64, 65, 63. Sixty-four is the smallest number, after 1, that is both a square and a cube ($64 = 8^2 = 4^3$). Sixty-five is the smallest number which is the sum of two squares in two different ways ($65 = 8^2 + 1^2 = 4^2 + 7^2$). Sixty-three is the smallest number which is not prime, not a cube, not 1 more than a multiple of 4, and not a multiple of the sum of two squares! (For example, 15 is a multiple of $5 = 2^2 + 1^2$; 21 is 1 more than 5×4; 27 is the cube of 3.)

In arithmetic these three numbers are lined up like soldiers on parade. Sixty-four is not allowed to flaunt its special traits but must keep step, in front of 65 and after 63. In fact, it is important, if you want to be good at arithmetic, to be able to handle all numbers with the same regular rhythm, just as a first baseman tries to cultivate the same rhythm for the hard throws as for the easy ones. So when we study the special properties of individual numbers, we are venturing far from arithmetic.

This venture, "useless" though it may seem, has attracted men for at least twenty-five centuries. In ancient Greek times, there was a school of numerology, headed by Pythagoras, which was more of a religious cult than an academic institution. The Pythagoreans were not only interested in numbers, they also believed that everything in the world is made out of numbers, so that if you understand numbers you can understand everything. Nowadays, of course, there are many fields of study that require a knowledge of mathematics, but the Pythagoreans meant something different. They thought that each number has its special

Pythagoras

qualities. For example, 4 signifies justice and 3 signifies power. They also believed that the qualities of a physical object are determined by the numbers that go into its construction. That is, a house built with 482 bricks would have the special qualities of 482. However, no one can be absolutely sure what the Pythagoreans believed, because they considered their ideas to be religious mysteries, which could not be revealed to outsiders. So they did not write any books about their system, and all we know of them comes from other Greeks who were not in the cult, but only wrote down what they had heard about it.

At any rate, the followers of Pythagoras probably had a table of what they considered to be the qualities of numbers. When a number is the smallest to possess some property, it becomes interesting. We must then say that unless we count as far as that number, we cannot know all there is to know about numbers. For example, if we had never counted as far as 64, we might think that it was impossible for any number besides 1 to be both a square and a cube. When we reached 64, we would see that we had been mistaken. Thus, 64 is an interesting number; it shows us a new possibility of numbers.

Now, you might think that most numbers are uninteresting, but one can easily be fooled. There is a story about the number 1729 and the brilliant Indian mathematician Ramanujan, who died in 1920. Once when Ramanujan was sick, a friend came to visit him in a taxi with the number 1729 on it. The friend could not think of anything interesting about this number, although he tried during the whole trip. When he arrived, he told Ramanujan about the uninteresting number. Ramanujan was famous for his intimate knowledge of numbers. He said immediately, "Why, 1729 is the smallest number which is the sum of two cubes in two different ways!" (One way he had in mind was $1729 = 1728 + 1 = 12^3 + 1^3$. Can you find the other way?)

But even Ramanujan might not be able to think of anything

interesting about, let us say, a six-digit number. There is a very clever argument, though, for the proposition that every number is interesting in some way. Suppose that you make a list of interesting numbers, and that it includes all numbers up to 113789, but that 113789 is uninteresting. Then 113789 is the smallest uninteresting number, which is very interesting—so 113789 should be included on the list. But then 113789 is no longer uninteresting, so that it cannot be called the smallest uninteresting number. But then it should *not* be on the list. This is a paradox, and it shows that there cannot be any smallest uninteresting number (for such a number would be interesting), and therefore there cannot be *any* uninteresting number.

Now, this argument shouldn't be taken too seriously, because it assumes that every number is either interesting or not. Actually there are degrees, and as you count higher the numbers tend to get less interesting. Just which numbers are more interesting, and which less, is something about which there is no absolute rule. But I should like to see Ramanujan's own table of interesting properties of the first thousand numbers.

After the death of Pythagoras, his followers split into two groups. One group was mainly interested in worshipping numbers, and the other in studying them. The more religious division didn't last long, and it is easy to see why. It depended too much on controversial things. For example, I have said before that to me the odd numbers appear darker than the even. But in the Chinese philosophy of Yang and Yin, the odd numbers represent light, and the even numbers, darkness. There is no way to prove that either opinion is right, and if any two students of Pythagoras disagreed about such things, they could only appeal to the master for a decision. Once he was dead, there was no way to settle differences of opinion, and the cult gradually broke up.

It was the other followers of Pythagoras, those interested in exploring numbers as they are, who began something which *did* last and is still going on. They discovered some real facts about numbers, which don't depend on anyone's opinion. These facts are called *theorems*, from a Greek word meaning "look." Let's examine the difference between a *theorem* and a *theory*, and between the scientific and the popular meaning of the word "theory."

Only a Theory

In mathematics, a theorem is a statement that can be proved. For example, a simple theorem about numbers is that the product of two consecutive numbers is always even. (Examples: $3 \times 4 = 12$, $8 \times 9 = 72$, $22 \times 23 = 506$.) The proof is as follows. Of two consecutive numbers, one is always odd and one even. The product of an odd and an even number is always even, because the even factor is "$2\times$" some other number. Thus, if m is odd and n is even, then $n = 2k$, and $mn = 2mk$. So mn is even, being twice mk.

The proof of a theorem is like a legal argument. When you write a proof, you must imagine that there is a lawyer working against you who claims that the theorem is not true and who will raise objections whenever he can. If we had stopped after asserting that the product of an odd and an even number is always even, the opposing lawyer might have said, "Maybe not." So we backed up the assertion with an argument.

A theorem always consists of a statement and a proof. Sometimes a theorem is stated without proof, but the proof must exist or the theorem is not a theorem. The statement without the proof is like a menu without the dinner. When you read the menu, you think about the items and build up an appetite. If no one comes to serve you the dinner, you may be hungry enough to cook it yourself. But a menu is not a recipe, and there is nothing in the statement of a theorem to advise you how to construct the proof. Once in the kitchen, you are on your own. Here are the statements of two theorems; the proofs will be "served" later in the book.

1. The square of any odd number is 1 more than some multiple of 8. (Examples: $5^2 = 1 + 8 \times 3$, $9^2 = 1 + 8 \times 10$.)

2. The sum of the first so many cubes is always a square. (Examples: $1^3 + 2^3 + 3^3 = 1 + 8 + 27 = 6^2$, $1^3 + 2^3 + 3^3 + 4^3 = 1 + 8 + 27 + 64 = 10^2$.)

In everyday language, a theory is just the opposite of a theorem. It is a statement that has *not* been proved and may even be false. Thus, Inspector Higginbottom has a theory that Spangley is the murderer, but he may be wrong. My wife has a theory that no man ever takes cream without sugar in his coffee, but she cannot prove it. In the popular sense, the word gives an impression of uncertainty and incompleteness; we say, "It is only a theory."

There is, however, a second meaning in the word theory. In this, the scientific meaning, a theory is not a statement at all, but a whole body of knowledge dealing with the ideas and rules behind some activity. If you have studied music and have taken a course in theory, you have learned about the relations of notes and chords and the rules of harmony. This is a very definite and precise subject, and there is nothing uncertain about it. Nor does it make sense to say that musical theory may or may not be true, because it is a whole subject and not a single statement. It is called "theory" because it is abstract and serves as background for the actual listening to or writing of music. In the same way, the "theory of perspective" is not a piece of guesswork but a definite body of rules and reasons which a draftsman (someone who draws) must master in order to draw pictures that give an impression of depth.

In science and mathematics, a statement that has not been proved is called not a theory, but an hypothesis. *Hypo* is the

Greek word for "under," and *thesis* for "put," which makes sense because *sub* is Latin for "under" and *pose* for "put." So an hypothesis is something you *suppose,* something you put at the bottom of your mind in order to see what can be built on it.

Sometimes the scientific and popular meanings are combined. The name theory is often given to a whole body of reasoning which rests on some unproven assumptions. Examples are Darwin's theory of evolution, Karl Marx's theory of capitalist economies, and Einstein's theory of relativity. This kind of theory resembles an hypothesis in that it can be believed or disbelieved, and evidence may be brought to support it. If enough evidence is brought against it, the theory may be overthrown, like the phlogiston theory of chemistry around 1800 or the caloric theory of heat around 1850. But a theory, in this sense, is more than an hypothesis, for it is not a mere statement but a whole body of reasoning. The theory of evolution is not just the assertion that animals have developed over many generations from other animals; it is the whole study of how they have developed, how fast, when, where, and why. Even Inspector Higginbottom's theory about the murderer is usually more than a bald hypothesis; it is a chain of reasoning that explains why Spangley wanted to kill his uncle, who the mysterious woman was, and how Spangley got out through the locked door.

The mixing of meanings sometimes generates false impressions about theories. The theory of relativity, when proposed in 1905, consisted mainly of a few bold hypotheses and some interesting consequences. It was appropriate, then, to say "It is only a theory," because the hypotheses were unsure. But, since then, the basic hypotheses and many striking consequences have been confirmed by experiment. (This does not make them theorems, because a theorem can be proved without experiment, just by thinking.) The theory of relativity is no longer uncertain; it is quite as firmly established as any other branch of physics.

A physicist studies the theory of relativity just as a radio engineer studies the theory of electric circuits. Calling it a theory does not mean that you doubt relativity, or doubt the existence of electric circuits. It just means that you have a lot to learn before you can build a radio. So it is no longer appropriate to say that the theory of relativity is "only a theory," as one might have said in 1910. Then it was a theory in the popular sense; now it is a theory in the scientific sense.

This book is about the "theory of numbers," which is a theory in the pure scientific sense, like musical theory or the theory of perspective. Number theory is composed mainly of theorems about numbers, together with the ideas that make the theorems interesting. There are also hypotheses in number theory, but the theory is not based on them. If such an hypothesis were disproved, the theory would not be overthrown, but furthered. The disproof itself would become part of number theory.

Dreaming of Mousetraps

Number theory is not mystical, like the numerology of Pythagoras. It is not concerned with whether odd numbers are dark or light, or whether 10 is a nobler number than 6. In number theory, we deal only with absolute truths which can be proved to everyone's satisfaction. In that way it is like arithmetic.

The difference between theory of numbers and arithmetic is like the difference between poetry and grammar. Poetry, like grammar, has rules, but it also has taste. Within the rules, a poem moves according to the poet's desire. Number theory obeys the same rules as arithmetic; in both disciplines, $2 + 2$ is always 4 and never 5. But number theory is motivated by its ancestor, numerology, in the choice of theorems. It concentrates on theorems that satisfy our inward curiosity about numbers, or give the illusion of deep mystery. This choice is subjective, and not all number theorists are fascinated by the same theorems. What distinguishes number theory is that its choice of material is governed by curiosity and wonder, whereas that of arithmetic is governed by use and convenience.

Some of the theorems in number theory (see Chapter 6) are useful in arithmetic, but most have no practical use at all. On the other hand, although the rules of arithmetic *could* be stated and proved as theorems in number theory, one has no desire to do so. For example, if we wrote out a set of instructions for long multiplication (including a tedious description of "carrying," and a prescription for putting the partial products in the right columns), we might then state and prove a theorem: this method of procedure always gives the correct product. But the mere statement of this theorem is so long and complicated that it does not arouse any natural interest. It is unbeautiful though useful, just as the telephone book is unpoetic though useful.

Consider, on the other hand, the simple theorem: The sum of two cubes is never a cube. (See Chapters 5 and 7.) There is no use for this fact, but one can easily become interested in it. The idea of a cube number, like $2^3 = 2 \times 2 \times 2 = 8$ or $3^3 = 3 \times 3 \times 3 = 27$, is a simple and familiar one. The operation of addition is simple and familiar. As soon as the idea is suggested, we are surprised not to know whether any cube is the sum of two cubes; we try to find such a cube, and are intrigued when we cannot. We wonder whether it is really true that none exists, and how it can be proved. The most exciting theorems of number theory are easy to state and hard to prove, just as the best mousetraps are easy to enter and hard to leave. A good theorem is a mousetrap for the mind.

The interesting number theorems are not only simple, they are universal. They are not suggested by the way we write numbers, or by anything in our way of life. In fact the strongest lure of numbers is in their permanence, their insistence on being what they are. Men may die and be born, may swim or fly, may wear skins or feathers. But the cube of 3 goes right on being just 2 more than the square of 5, whether or not anyone is looking.

After one has studied number theory, one feels that arithmetic is not about numbers at all, only about the names of numbers—their shadows, one might say. We ask, what is $18 \times 13 + 5$? The student of arithmetic computes for a minute and replies, "It is 239." But the number theorist says, "You have only told us its name. What is its nature?" Soldiers, when captured by the enemy, tell only their name, rank, and serial number. They tell nothing important, such as the kind of unit they were with or the direction in which it was moving. This number, 239, on being taken prisoner by the arithmetician, divulged only its name and rank, "I am the number after 238 and before 240, sir!" But the number theorist strikes up a closer acquaintance and soon learns intimate details about the soldier's family, his likes and dislikes—in short, his personality. He learns that 239 is a prime number, and one of a pair of *twin primes,* 239 and 241. Twin primes are fairly common; there are 36 such pairs below 1000. Two hundred and thirty-nine is not the sum of two squares, nor is it the sum of a square and double another square. In fact, the smallest number which can be multiplied by a square, and the product added to another square to make 239, is 7 ($239 = 64 + 7 \times 25$). What is more, 7 is the only solution if we do not allow it to be multiplied by 1^2, as in the equation $239 = 225 + 14 \times 1$. Can you find any other prime number besides 239 of which this is true?

Carl Friedrich Gauss (1777–1855), one of the founders of modern number theory, called it the "higher arithmetic." It is certainly a far cry from the wispy dreams with which this chapter began. We outgrow such fancies because they do not satisfy the need to build something that lasts. And yet the theory of numbers owes its superiority over "common arithmetic" to the dreams of its origin, which linger close enough to guide it into the paths of delight. To build upon dreams is what we aspire to most of all.

2.

On a clear day you can count forever

Guess the Pattern

In an intelligence test, you may be asked to continue a sequence by guessing a pattern. For example, find the next number in the following sequences:

(a) 1, 3, 5, 7, . . .
(b) 2, 1, 4, 3, 6, . . .
(c) 1, 2, 3, 5, . . .

The answer to (a) is easy. Each number is 2 more than the one before it. The next number is 9. Or you may say that the sequence contains all the odd numbers, in order. This answer is the same as the first one, for either way the sequence continues 9, 11, 13, 15, and so on forever.

Sequence (b) is about as hard as the ones given on intelligence tests. Following it from one number to the next, we find that it goes down 1, up 3, down 1, up 3. It is due to go down 1, so that the number after 6 should be 5.

Sequence (c) would not appear on a test—not because it is too hard, but because there are several different answers and no way to choose among them. A clever answer is that (c) is a list of all prime numbers. Then the next number is 7.

(c_1) 1, 2, 3, 5, 7, 11, 13, 17, . . .

But perhaps it is a list of *all* numbers, leaving out every fourth one; then it must continue with 6.

(c_2) 1, 2, 3, 5, 6, 7, 9, 10, 11, . . .

Again, you may notice that 3 is the sum of the two numbers before it, 1 and 2, and that 5 is the sum of 2 and 3. According to this pattern, the next number should be $3 + 5 = 8$, and the one after should be $5 + 8 = 13$.

(c_3) 1, 2, 3, 5, 8, 13, 21 . . .

Even if you are told that the next number is 7, you cannot be sure that (c) is a list of all primes. Maybe the sequence goes

(c_4) 1, 2, 3, 5, 7, 10, 13, 17, 21, etc.

with two jumps of 1, two jumps of 2, two jumps of 3, and so on.

The matter is still not settled if I say that the number after 7 is 11. Then (c) *may* be the list of primes (c_1) or it may proceed by pairs of jumps each twice as big as the previous pair, thus

(c_5) 1, 2, 3, 5, 7, 11, 15, 23, 31, etc.

All this goes to show that you cannot really tell what a sequence is by knowing the first four or five numbers—or even by knowing the first five hundred numbers. There are always many patterns to choose from. In a test, one pattern is much more simple and obvious than any other, and you are supposed to choose that one. But you can always find others if you try. For example, sequence (a) might go

(a_1) 1, 3, 5, 7, 8, 6, 4, 2, 3, 5, 7, 9, 10, etc.

To describe a sequence completely, one must give a rule for finding all the numbers in it. There are different kinds of rules, and I shall call three of them *progressive, ordinal,* and *inclusive.*

A *progressive* rule is one that tells how to go from each number in the sequence to the next. In sequence (a), we add 2 to the previous number. In sequence (b), we alternately subtract 1 and add 3. In (c_3) we add the last number to the one before it to get the next one. This is also a progressive description, although it refers to two previous numbers instead of just the last one.

An *ordinal* rule tells how to find the number that comes in any given order in the sequence. That is, it tells you how to find the tenth, or the twenty-sixth, or the forty-second number, without counting through the whole sequence as you would have to do with a progressive rule.

An ordinal rule for sequence (a) is that the number in position m is $2m - 1$. Thus, the number in position 1 is $2 \times 1 - 1$, or 1. The number in position two is $2 \times 2 - 1$, or 3. The third number is $2 \times 3 - 1$, or 5. By this rule you can find out that the forty-first number is $2 \times 41 - 1$, or 81, much faster than by adding 2 over and over.

It is often possible, given an ordinal rule, to find a progressive rule for the same sequence. One sequence with a simple ordinal rule is the list of squares

(d) 1, 4, 9, 16, 25, 36, . . .

The ordinal rule is that the number in position m (we may call it the "mth" number) is $m \times m$, or m^2. The third number is 3×3, and so on. But there is a progressive rule that tells us to add 3, then add 5, then add 7, and so on. Another progressive rule which amounts to the same thing is: add 1 to the last number, double it, and subtract the next-to-last. If you add 1 to 36, double, and subtract 25, you get 49, which is the next square number after 36.

But you cannot always find a progressive rule. Here is an ordinal rule: the mth number is the number of factors that m has, not counting itself. The number 8 has three factors (1, 2, and 4), and so the eighth number will be 3. The sequence goes

(e) 0, 1, 1, 2, 1, 3, 1, 3, 2, 3, 1, 5, 1, 3, 3, 4, 1, 5, 1, 5, . . .

Just try and find a progressive rule for that!

The sequence (c_3) has a progressive rule, as I mentioned. I don't know any ordinal rule for it. That is, I don't know how to find the twentieth number without working out all the numbers before it in the sequence.*

Serpents and Subways

An *inclusive* rule is one that tells which numbers are in the sequence and which are not. Sequence (a) contains all odd numbers; that is an inclusive rule. If the numbers are out of order, as they are in sequence (b), an inclusive rule does not tell what order they are in. The inclusive rule for (b) would say that it contains all the numbers, which is true but misses the point.

Even if we know the numbers are in order, it may be difficult to go from an inclusive to a progressive or ordinal rule. The sequence (c_1) is described inclusively as the list of all prime numbers, but neither an ordinal nor a progressive rule can be found. There is no rule for finding the twentieth prime number after 47—except by trying out each number to see whether it is prime.

You might expect that it is easier to find an inclusive rule, since the inclusive rule doesn't have to put the numbers in any order. But sometimes it is quite hard. Here is a sequence with an ordinal rule:

(f) 5, 17, 37, 5, 101, 5, 197, 257, 5, 401, 5, 577, 677, 5, 17, 5, 13, 1297, 5, 1601 . . .

The mth number is the smallest factor of $4m^2 + 1$, not counting 1 as a factor. If m is 6, then m^2 is 36, and $4m^2 + 1$ is $144 + 1$, or 145, and the smallest factor of 145 is 5. So the sixth number is 5. If m is 7, then $4m^2 + 1$ comes out to be 197, and nothing goes into 197, besides 1, except 197. So the seventh number is 197.

* See note in Appendix 1.

An inclusive rule would tell us whether any particular number, such as 3, is in the sequence or not. You cannot find this out just by using the ordinal rule and working out the sequence, because of the jumps back and forth. It seems that 3 has been skipped, but maybe it will turn up later. The number 13 turned up for the first time in position seventeen, after some much larger numbers. Even if you find a thousand numbers in the sequence, with no 3, you won't know whether a 3 will turn up later.

To find an inclusive rule for (f) requires deep thought. A little reflection shows that nothing besides prime numbers can be in the list, and that 2 is not in the list. If you work hard, you may see a way to prove that 3 is not in the list either. (If you can do that on your own, without knowing any number theory before, you have unusual ability.) But to find the rule that is good for all numbers is especially difficult, and we shall work on it in Chapter 7.

There are sequences with an ordinal rule but no inclusive rule. I can't describe them here, except in a general way. Imagine that a sequence could read your mind. As soon as you thought of an inclusive rule, the sequence would turn back, like an ill-tempered serpent, and include one of the very numbers that your rule said should not be included. Or if your rule included a lot of numbers, the sequence would disobey it by deliberately avoiding one of those numbers, forever and ever. Now it is possible to describe, by a very complicated ordinal rule, just such a "mind-reading" sequence. What the ordinal rule actually does is to figure out all the inclusive rules that anyone could ever imagine, one after the other, and make the sequence disobey every one. So there cannot be any inclusive rule for this sequence. This sort of mind-reading sequence belongs to a subject called *recursion theory,* which began only in the 1930's. But recursion theory is not number theory, and we have a long trip ahead.

Before going ahead with the ordinal, progressive, and inclusive rules, let me mention a fourth type which I shall call *extraneous.* This is a rule that connects the numbers with something in your life that has nothing to do with numbers. Extraneous rules are not scientific and do not belong in number theory, but they make good puzzles. I remember one that circulated among my friends in college. The sequence is:

(g) 14, 23, 28, 33, 42, 49, 59, 66, 77, 86, 96, 103, . . .

This was good for a wasted hour. The numbers seem to go up regularly, but the jumps keep changing—sometimes 7, sometimes 9 or 5. The numbers do seem to alternate, odd and even—but no, there is 49 followed by 59. How about factors? None of them are multiples of 5; is that a clue? No, it isn't. The number after 103 is 110. These numbers are the streets at which the Lexington Avenue subway stops, in New York City!

Try your hand on these sequences. Each has an extraneous rule. (Answers upside-down at bottom of page 31.)

(g_1) 7, 5, 15, 18, 7, 5, 0, 23, 1, 19, 8, 9, 14, 7, 20, 15, 14.
(g_2) 5, 5, 5, 3, 4, 4, 4, 2, 5, 5, 5, 3, 6, 6, 6, 5, 10, 10, 10, 8, . . .
(g_3) 76, 81, 12, 15, 45, 48, 61, 65, 98, 0, 17, 18, 41, 45, 50, 52, 65, 73, 91, 91, . . .
(g_4) 1, 4, 9, 61, 52, 63, 94, 46, 18, 1, 121, 441, 961, 691, . . .
(g_5) 0, 1, 8, 11, 69, 88, 96, 101, 111, 181, 609, 619, 689, 808, 818, 888, 906, 916, 986, 1001, 1111, . . .
(g_6) OTTFFSSEN . . .

Hard Sums Made Easy

The simplest kind of progressive rule is the kind that tells you what to add to each number to get the next one. A sequence with this kind of rule given is called a *series,* and the amounts to be added at each step are the *terms* of the series. A series is usually written down by putting its terms with addition signs between them. For example, the first two sequences in this chapter can be written as series

(a) $1 + 2 + 2 + 2 \ldots$
(b) $2 - 1 + 3 - 1 + 3 - 1 \ldots$

The terms of (a) are 1, 2, 2, etc. The terms of (b) are 2, $-$ 1, 3, $-$ 1, 3, $-$ 1, etc. It is important to understand that the *series* $1 + 2 + 2 + \ldots$ is the same as the *sequence* 1, 3, 5, . . . , and different from the *sequence* 1, 2, 2, ...

To "sum" a series is to find an ordinal rule for the sequence it represents. For example, we can sum the series

(h) $1 + 3 + 5 + 7 + \ldots$

by saying that the sum of the first n odd numbers is n^2. An easy way to understand this is by an example. Let $n = 4$. Then we are adding $1 + 3 + 5 + 7$. Now some of the terms are less than 4 and some are greater. In fact, 4 is halfway between 7 and 1 and also between 5 and 3. So $1 + 3 + 5 + 7 = 4 + 4 + 4 + 4 = 4^2$.

In the same way, $1+3+5+7+9=5+5+5+5+5=5^2$ (note that $1 + 9 = 5 + 5$, etc.), and in general the sum of the first n odd numbers is n^2.

ANSWERS

(g1) The name of a President; 1 stands for A, 2 for B, etc.
(g2) Beethoven's Fifth Symphony; 1 stands for the first note in the minor scale, etc.
(g3) The first and last years of U.S. wars. The Revolution began in '76 and ended in '81.
(g4) The squares of numbers, written backward; 16 backward is 61.
(g5) Numbers that look the same when turned upside-down.
(g6) Ask your friends, someone will guess it.

We can use the same trick to sum the series

$$\text{(i)} \qquad 1 + 2 + 3 + 4 + \ldots$$

This time, if we take the first 4 numbers, the average is $2\frac{1}{2}$. That is, $1 + 4 = 2\frac{1}{2} + 2\frac{1}{2}$. If we take the first n numbers, the average is $\frac{1}{2}(n + 1)$, which means half of $n + 1$. So the sum is n times that, or $\frac{1}{2}n(n + 1)$.

Objection: How can the sum of whole numbers be a fraction? Answer: $\frac{1}{2}n(n + 1)$ is not a fraction, because $n(n + 1)$ is always an even number, whatever n is. (Try it and see. In fact, this was proved in Chapter 1.)

Let us try the series

$$\text{(j)} \qquad 1 + 4 + 9 + 16 + \ldots$$

In other words, find a formula for the sum of the first n squares. The averaging trick won't work here because $1 + 16$ is not the same as $4 + 9$, etc. So let us feel our way. The first few sums are

$$1^2 = 1$$
$$1^2 + 2^2 = 5$$
$$1^2 + 2^2 + 3^2 = 14$$
$$1^2 + 2^2 + 3^2 + 4^2 = 30$$
$$1^2 + 2^2 + 3^2 + 4^2 + 5^2 = 55 = 11 \times 5$$
$$1^2 + 2^2 + 3^2 + 4^2 + 5^2 + 6^2 = 91 = 13 \times 7$$

We notice that the fifth sum is a multiple of 11 and the sixth sum is a multiple of 13. From this we might guess that the seventh sum will be a multiple of 15, the eighth of 17, and so on. Let's try it.

$$1^2 + 2^2 + 3^2 + 4^2 + 5^2 + 6^2 + 7^2 = 140 = 15 \times 9\frac{1}{3}$$
$$1^2 + 2^2 + 3^2 + 4^2 + 5^2 + 6^2 + 7^2 + 8^2 = 204 = 17 \times 12$$

It works for the eighth sum but not for the seventh. Looking back, we see that the first and fourth sums don't work either (1 is not a multiple of 3, nor 30 of 9) but that all the rest do. We shouldn't give up here because it must mean something if a pattern is followed most of the time, even if it doesn't work perfectly. For example, we know now that there is a good chance that the ninth sum will be a multiple of 19. And it is.

$$1^2 + 2^2 + 3^2 + 4^2 + 5^2 + 6^2 + 7^2 + 8^2 + 9^2 = 285 = 19 \times 15$$

Without looking at the earlier sums, we should never have guessed this in advance. We must have stumbled on something; but why doesn't it always work?

Look at the cases that don't work. They yield fractions where they ought to yield whole numbers. That is, $1 = 3 \times \frac{1}{3}$, $30 = 9 \times 3\frac{1}{3}$, $140 = 15 \times 9\frac{1}{3}$. But the fraction is always in thirds. So if we multiply the sum of squares by 3, all these fractions become whole numbers. Now we have:

$$\begin{aligned}
3 \times 1 &= 3 = 3 \times 1\\
3 \times 5 &= 15 = 5 \times 3\\
3 \times 14 &= 42 = 7 \times 6\\
3 \times 30 &= 90 = 9 \times 10\\
3 \times 55 &= 165 = 11 \times 15\\
3 \times 91 &= 273 = 13 \times 21\\
3 \times 140 &= 420 = 15 \times 28\\
3 \times 204 &= 612 = 17 \times 36\\
3 \times 285 &= 855 = 19 \times 45
\end{aligned}$$

Suddenly it all turns clear. The numbers 3, 5, 7, . . . go up by twos, and the numbers 1, 3, 6, 10, . . . go up by increasing jumps. In fact, they are just the sums $1 + 2 + 3 + 4 + \ldots$. For the tenth row in the table, we expect to increase 19 by 2 and increase 45 by 10. So the sum of the first ten squares, multiplied by 3, should give 21×55. Let's try it.

$$3 \times (1^2 + 2^2 + 3^2 + 4^2 + 5^2 + 6^2 + 7^2 + 8^2 + 9^2 + 10^2)$$

$$= 3 \times 385 = 1155 = 21 \times 55$$

Hooray!

Since we already have ordinal rules for the sequences 3, 5, 7, . . . (the nth number is $2n + 1$) and 1, 3, 6, . . . (the nth number is $\frac{1}{2}n(n + 1)$, as we figured out before), we can write down the formula for the sum of the first n squares

$$1^2 + 2^2 + \ldots n^2 = \tfrac{1}{3} \times (2n + 1) \times \tfrac{1}{2}n(n + 1)$$
$$= \tfrac{1}{6}n(n + 1)(2n + 1)$$

Let us try the formula once more, for $n = 11$.

$$1^2 + 2^2 + 3^2 + 4^2 + 5^2 + 6^2 + 7^2 + 8^2 + 9^2 + 10^2 + 11^2 = 506$$

$$\tfrac{1}{6} \times 11 \times (11 + 1) \times (2 \times 11 + 1) = \tfrac{1}{6} \times 11 \times 12 \times 23$$
$$= 11 \times 2 \times 23 = 506$$

You may enjoy trying the same methods to find a formula for the sums

$1 = 1$
$1 + 3 = 4$
$1 + 3 + 6 = 10$
$1 + 3 + 6 + 10 = 20$
$1 + 3 + 6 + 10 + 15 = 35$, and so on.

Taking the Cure

But hold on! How do we know that the formula $\frac{1}{6}n(n + 1)(2n + 1)$ for the sum of the first n squares, is *always* right? We haven't proved it. We have only shown that it works for $n = 1$, $n = 2$, and so on up to $n = 11$. Maybe it will be wrong for $n = 12$. If it works for $n = 12$, maybe it won't for $n = 13$.

"Oh," you may say, "anything that works eleven times out

of eleven must be right." To cure yourself of this idea, consider the sequence

(k) 41, 43, 47, 53, 61, 71, 83, 97, 113, 131, . . .

It can be written as a series

(k) $41 + 2 + 4 + 6 + 8 + 10 + 12 + 14 + 16 + 18 + \dots$

Now, 41 is a prime number. So are 43, 47, 53, and all the rest in the sequence up to 131. Does that mean that the next number will be prime? The next is 151; it is prime. So is 173, which comes after 151. The next few are 197, 223, 251. All these are prime numbers. (Try finding a factor for any of them.) In fact, you can go on until you are tired without reaching a number in this sequence which is not prime.

Does this mean that *every* number in the sequence is prime? No. The forty-first number is not prime. (You will be tired before you reach it.) This number is $41 + 2 + 4 + 6 + \cdots + 80$. The dots mean that we must add all the even numbers up to 80. Now this sum can be done by averages: $2 + 80 = 41 + 41$; $4 + 78 = 41 + 41$, and so on. So the number in question is just $41 + 41 + 41 + \cdots + 41 = 41 \times 41$, which is not prime.

Not cured yet? Try this one. The sequence

(l) 2, 4, 16, 256, 65536, . . .

is made by squaring each number to get the next one. If we add 1 to each number in the sequence we get

(m) 3, 5, 17, 257, 65537, . . .

These are called "Fermat numbers" because Pierre Fermat, a very clever Frenchman of the seventeenth century, took an interest in them. He guessed that they would all be prime. It is easy to see that 3, 5, 17, and 257 are prime, but how about 65537? It looks hard, but isn't, really. It can be checked in a day, by a method called the "sieve of Eratosthenes" (see

Chapter 3). It can also be done in ten minutes by more advanced methods (see Chapter 6). The sieve of Eratosthenes was well known in Fermat's time. The more advanced methods were not, but that does not mean that Fermat didn't know them. No one is sure how much Fermat knew. This is not because he belonged to a secret society like the cult of Pythagoras, but because he never bothered to tell anyone. He was a lawyer who studied number theory for fun, and solved problems that people wrote to him about. When he had solved one, he would write his friend a letter giving the answer, but he hardly ever explained how he had found it. He must have known tricks that no one else knew, because many of his answers were found to be correct long after he died, when mathematics had advanced enough so that the problems could be solved.

At any rate, Fermat was certainly capable of finding that 65537 is a prime number. But the next Fermat number was beyond him. It is $65536^2 + 1$, which is a number of ten digits. Fermat died without knowing for sure whether *that* number is prime. But he guessed that since the numbers 3, 5, 17, 257, 65537 are all prime, the next number in the sequence would be also.

He guessed wrong. The next Fermat number is not prime. (See Chapter 6.) In fact, it is now known that the next *few* Fermat numbers are not prime. So now we are tempted to guess that *none* of the Fermat numbers, after 65537, are prime. But we resist the temptation, because by now we are cured of guessing!

There is a famous guess that still plagues everyone. It is called Goldbach's postulate, and it says that every even number is the sum of two primes. For example, $2 = 1 + 1, 4 = 1 + 3, 6 = 1 + 5$, $8 = 1 + 7$. We won't write $10 = 1 + 9$, since 9 isn't prime, but we can use $10 = 3 + 7$. Another way to state Goldbach's postulate is to say that if you roll two dice, each of which has lots of faces (instead of only six), and each face has a prime number of pips (that is, each die has a 1 face, a 2 face, a 3 face, a 5 face, a 7 face, an 11 face, a 13 face, and so on, but no 6 face or 9 face) then you can roll any even number as the total.

Goldbach's postulate is very easy to check (until you get tired) and very hard to prove. It has not been either proven or disproven to this day, although it is known to be true for all even numbers up to 10,000 and many beyond.

Mathematical Induction

We may now take a warier attitude toward our lovely formula, $\frac{1}{6}n(n+1)(2n+1)$, for the sum of the first n squares. Perhaps it is not always true. Let us examine it.

Suppose we use the formula to find the sum of the first $(n-1)$ squares. Then we must take one-sixth of $(n-1)$, times one more than $(n-1)$, times one more than twice $(n-1)$. This comes out $\frac{1}{6}(n-1)n(2n-1)$. But if this formula is right, we can add n^2 to it to get the sum of the first n squares. Then we have $\frac{1}{6}(n-1)n(2n-1)+n^2$. Can this be the same as $\frac{1}{6}(n+1)(2n+1)$?

If you know even a little algebra, you can multiply $(n+1)$ by $(2n+1)$ and get $2n^2+3n+1$. Therefore $\frac{1}{6}n(n+1)(2n+1)=\frac{1}{3}n^3+\frac{1}{2}n^2+\frac{1}{6}n$. In the same way you can find that $\frac{1}{6}(n-1)n(2n-1)=\frac{1}{3}n^3-\frac{1}{2}n^2+\frac{1}{6}$. Adding n^2 to this, we have $\frac{1}{3}n^3+\frac{1}{2}n^2+\frac{1}{6}n$, which *is* the same as $\frac{1}{6}n(n+1)(2n+1)$. So the formula agrees with itself, so to speak. If it works for $n-1$, then it will work for n.

Now, agreeing with oneself has only a limited value. If the police check a man's alibi and find that he hasn't contradicted himself, it doesn't prove that he is telling the truth, although it makes it easier for them to believe him. But suppose they check one part of his story independently and find that it is true beyond a doubt. They may then deduce that if he is telling the truth about that part, then he couldn't be lying about the phone call he made. And if he really made the phone call, then he must be telling the truth about the note he found under his door, and if he wasn't lying about the note, then they have to believe his story about going to the movies, and so on, until his whole alibi is proven.

That is our situation with respect to the formula $\frac{1}{6}n(n + 1)(2n + 1)$. We have checked it for $n = 1$, $n = 2$, and so on up to $n = 11$. But now we know that if it works for $n - 1$, it will work for n. Let $n = 12$; then $n - 1 = 11$. We have checked the formula for 11; therefore we know without checking that it works for 12. Knowing this, we can let $n = 13$, $n - 1 = 12$. Since the formula works for 12, it must work for 13. By continuing this reasoning, we can prove that the formula works for 14, 15, and so on, up to 98 or 1267 or any number, no matter how large. But since we know now that the formula can be proved true for any number n, we know that it *is* true for all n. That is, we have proved it.

This method of proof is called *mathematical induction*. According to this method, to prove that a statement is true of all numbers you need only prove two things: first, the statement is true of the number 1, and second, if it is true of $n - 1$, then it will be true of n.

Let us try induction on another series

$$\text{(n)} \qquad 1 + 2 + 4 + 8 + \cdots$$

The nth term of this series is 2^{n-1}. (For example, the third term is 2^2.) The sums form the sequence

$$\text{(n)} \qquad 1, 3, 7, 15, \ldots$$

What is an ordinal rule for this sequence?

Mathematical induction won't help us guess the rule. It is

only a way of proving the rule once we have guessed it. Let us be observant. If we continue sequence (n), it goes

$$31, 63, 127, 255, \ldots$$

We notice that each number is *approximately* twice the one before it. This gives us the clever idea of adding 1 to each number. Then each number is *exactly* twice the one before it; in fact, they are all powers of 2. We have found an ordinal rule for the sum of the first n terms of (n),

$$1 + 2 + 2^2 + \cdots + 2^{n-1} = 2^n - 1$$

For example, if $n = 5$

$$1 + 2 + 2^2 + 2^3 + 2^4 = 2^5 - 1 = 31$$

Now, let us prove that the rule is always right. First step: It is right when $n = 1$. Just try it. The rule says that $1 = 2^1 - 1$, which is true since $2^1 = 2$.

Second step: Suppose the rule works for $n - 1$. That is, suppose that

$$1 + 2 + 2^2 + \cdots + 2^{n-2} = 2^{n-1} - 1$$

Adding 2^{n-1} to both sides

$$\begin{aligned} 1 + 2 + 2^2 + \cdots + 2^{n-2} + 2^{n-1} &= 2^{n-1} + 2^{n-1} - 1 \\ &= 2 \times 2^{n-1} - 1 \\ &= 2^n - 1 \end{aligned}$$

But this is just what the rule says, when applied to n. Therefore if the rule works for $n - 1$, it works for n.

Having completed the two steps of induction, we have proved the theorem

$$1 + 2 + 2^2 + \cdots 2^{n-1} = 2^n - 1, \qquad \text{for all } n$$

Try induction yourself to prove that the sum of the first n cubes is the square of $\frac{1}{2}n(n + 1)$ (see Chapter 1). You will have to show first that $[\frac{1}{2}n(n + 1)]^2 = [\frac{1}{2}(n - 1)n]^2 + n^3$.

People who hear of mathematical induction for the first time are often suspicious of it. They think it is a form of "circular reasoning." Circular reasoning is when you assume the very thing you are trying to prove. It is like pulling yourself up by your bootstraps. For example, "That man is a thief." "How do you know?" "He stole my jacket." "Did you see him steal it?" "No, but I see him wearing it." "He says he bought it in a store." "He's lying." "How do you know he's lying?" "Because he can't be trusted." "Why can't he be trusted?" "Because he's a thief."

This argument doesn't prove that the man is a thief, because it depends on knowing already that he is one. If we aren't sure of that already, then we can't be sure that he isn't to be trusted, that he's lying, that he didn't buy the jacket, and so on. And so we are no more certain that he is a thief at the end of the "proof" than we were at the beginning.

When you prove something by mathematical induction, you begin the second step by saying, "Suppose it is true for $n - 1$." At that point someone says, "Aha! That is exactly what you are trying to prove! You are using circular reasoning."

He is wrong. Mathematical induction does not go in a circle. It goes in a line, from 1 to 2 to 3 and so on up, never returning. If you said, "Suppose it is true for 5, then it will be true for 6," and followed that up with, "Suppose it is true for 6, then I can prove it for 5," you would be using circular reasoning. For 5 depends on 6, and 6 on 5, and you might be wrong about both of them. But in mathematical induction, 6 depends on 5, which depends on 4, which depends on 3, which depends on 2, which depends on 1, which is proved directly. This is not circular.

Birthdays and Bootstraps

I remember one trick proof which leads to an impossible

conclusion. The proof has a mistake, of course, but it is hard to find. It is a proof by induction, and that makes it especially tricky.

The "theorem" is that if any number of people are gathered in a room, they will all have the same birthday. Now this is obviously untrue, so that anyone who "proves" it must have an ace up his sleeve. Here is the proof.

We make it into a statement about the number n: If any n people are together in a room, they all have the same birthday. Then we use induction. First we prove it for $n = 1$. In other words, we prove that if any one person is in a room alone, then everyone in the room has the same birthday. This is so obvious that we can consider it proven.

Second, we assume the statement is true for $n - 1$, and prove it for n. Suppose there are n persons in a room. Call the last person who entered, Jones. Before Jones came in, there were only $n - 1$ persons. So those $n - 1$ must all have the same birthday. Jones is the only one who may be out of step. Now let Jones stay in the room and send someone else out—call him Smith. There are now $n - 1$ left in the room, including Jones. So Jones must have the same birthday as the others. Therefore all the original n, including Jones and Smith, have the same birthday.

We have proved the statement for $n = 1$, and we have proved that it is true for n if it is true for $n - 1$. Therefore we have proved that all people have the same birthday. What went wrong?

Most people's first reaction is that the method of induction is false. "You can't assume that $n - 1$ people have the same birthday. You haven't proved it." After you explain mathematical induction very carefully, they will say you have used it wrongly. "All right, you can assume that the $n - 1$ people in the room before Jones came in all had the same birthday. But after Smith left, there was a different group of $n - 1$. You don't know it's so for that group." But that is not a good objection, because the statement is phrased "If *any* n people, etc.," that is, not just one group of n people, but every group that can be brought together. In the second step of induction, we prove it for *every* group of n people. Therefore we can assume it for *every* group of $n - 1$ people.

When you have made this clear to your victim, he may say desperately, "Well, I don't believe you proved it for $n = 1$!" Now he is at the end of his rope, because the statement obviously *is* true for $n = 1$. Common sense tells us that.

The real flaw has nothing to do with the use of induction. It is a hidden assumption in the reasoning of the "second step." When Smith left the room, we observed that Jones must have the same birthday as the others who remained. But since we already knew that those others have the same birthday as Smith, we concluded that Smith and Jones have the same birthday. The trouble with this is that there may not be any "others." The argument is correct if $n = 3$ or more, but if $n = 2$, Jones is all alone after Smith leaves, just as Smith was alone before Jones came in. It is true that Jones has the same birthday as himself, but that does not mean he has the same birthday as Smith. The proof depends on the assumption that there is at least one other person in the room, first with Smith and then with Jones. And this is simply untrue if $n = 2$.

Since the proof breaks down for $n = 2$, the theorem has not been proved for any higher n. We have only proved that *if* it were true for $n = 2$, it would be true for all higher n. And this is so. If it were true that *any* two people have the same birthday (that is, that no two people have different birthdays) then it would be true that all people have the same birthday. In fact, the false proof I gave above is a perfectly good proof provided the statement is known to be true for $n = 2$. If you read it over in that light you will find that it makes sense and that you will not feel any objections to the use of induction but will regard it as quite obvious.

The false objections that people tend to raise to this tricky paradox are an example of a common pitfall in everyday life. When someone presents you with an argument that leads to a conclusion you know or feel is wrong, you will automatically become suspicious of every part of the argument. This can be costly, because probably there is only one part that is wrong and the rest is all right. If you do not spot the flaw right away, you may blunder hastily into the sound parts of the argument, raising absurd objections and feeling angry when they are answered. It is worth taking the time to tell yourself that if you are sure the conclusion is wrong then you must have a good reason, and if you examine that reason calmly you will see what is wrong with the argument.

You may also find yourself on the other side. If you are trying to explain something that you think is obvious and can't understand why the other fellow is objecting to everything you say, the chances are that he is afraid you are leading up to some conclusion that he won't accept. If you can guess what he thinks you are leading to, and assure him that you aren't, he will listen much better to what you *are* saying.

Two Theorems

I shall finish this chapter with two theorems that were discovered and proved by the ancient Greek mathematicians. The first had to do with the effort of Pythagoras to express everything by numbers. One thing Pythagoras wanted to do was express geometrical shapes by numbers—which, for him, meant whole numbers. For example, he wanted to find two whole numbers a and b such that, if he laid out a square with a units on each side, the diagonal would be b units long. Obviously, small numbers don't work; if the side of the square is 3 units long, the diagonal is between 4 and 5, and its length isn't a whole number.

Now, if you make a *bigger* square, built on the diagonal of the first one, it is easy to see that the second square has exactly twice

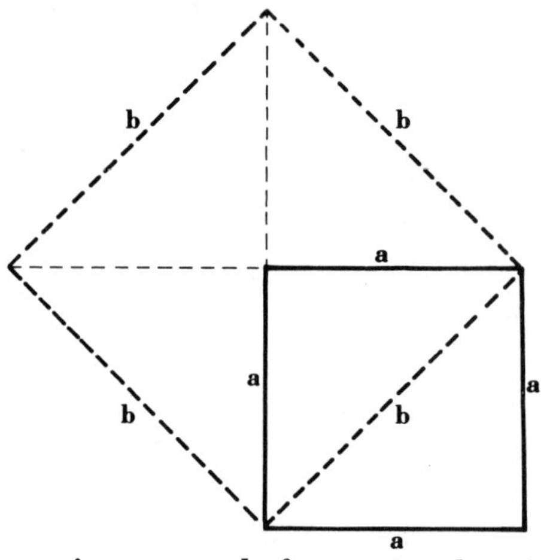

The a-square is composed of two triangles, the b-square of four; hence, $b^2 = 2a^2$.

the area of the first. So the problem is to find two square *numbers*, of which one is twice the other. The Pythagoreans eventually discovered that this is impossible, and it upset them no end.

The proof is simple. Suppose that $b^2 = 2a^2$. Then b^2 is an even number. That means that b is even, since the square of an odd number is always odd. (This is a familiar fact, but you might enjoy proving it by induction.) So half of b is a whole number, which we may call c. Then $b = 2c$, $b^2 = 4c^2$, and $a^2 = ½b^2 = 2c^2$. But then a must be even, and we can let d be half of a. In this way we can make smaller and smaller numbers by dividing by 2, without ever reaching an odd number. At each point we can prove that the numbers we have reached are even. So we can go on dividing forever. But this is impossible. Eventually we must reach 1, or some other odd number. Since we have proved something that is impossible, we must have assumed something that isn't true. The only thing we assumed was that there are two numbers a and b, such that $2a^2 = b$. So there can be no such numbers.

This is really a second form of induction, based on the idea that there obviously cannot be an infinite sequence of numbers going *downward*. If you find a way, given a number of a certain type, always to produce a *smaller* number of the same type, it proves that there is *no* number of that type. This is sometimes called *the method of infinite descent.*

The second theorem from ancient Greece is found in Euclid's *Elements*. This famous book is usually thought of as the foundation of geometry, but it is much more than that. Euclid lived more than two hundred years after Pythagoras. During those two hundred years, several brilliant mathematicians lived and died. The *Elements* is a carefully organized summary of the discoveries they made, some very important and original. Besides what we now call high-school geometry, the *Elements* contained what was known of number theory in Euclid's day, as well as the beginnings of some modern branches of mathematics.

The theorem I have in mind says that there is no end to the sequence of primes. The proof is supposed to have been discovered by Euclid himself. Given any number n there must be a prime number greater than n. To find it, multiply together all the numbers up to n. (This product is called $n!$; that is, $3! = 1 \times 2 \times 3 = 6$.) Now add 1. The number $n! + 1$ is not divisible by any number up to n; there is always just 1 left over. Now either $n! + 1$ is prime, or it has factors. If it has factors, the factors can be

Euclid

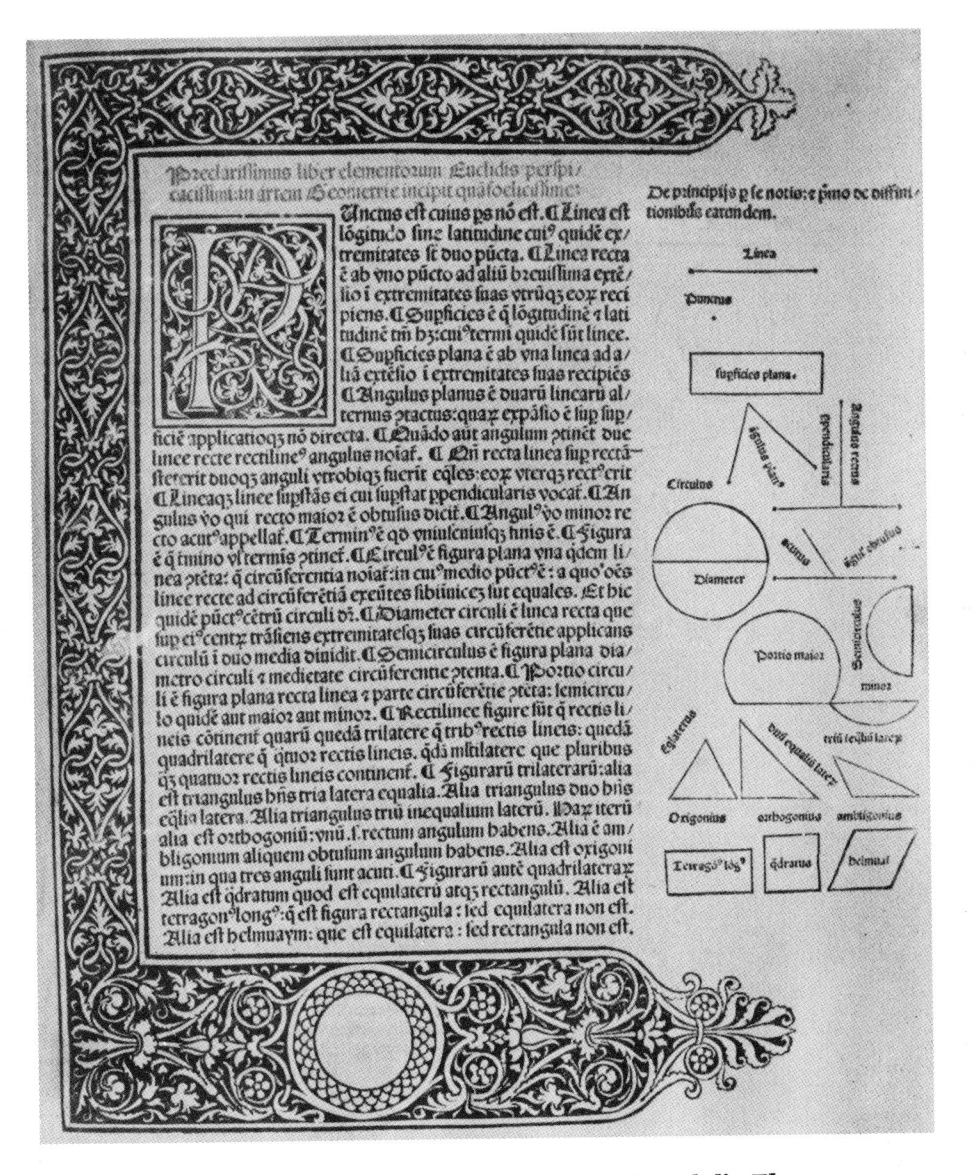

Preclarissimus liber elementorum Euclidis perspi/
cacissimi: in artem Geometrie incipit quã foelicissime:

Punctus est cuius ps nõ est. ¶ Linea est
lõgitudo sine latitudine cuiꝰ quidẽ ex/
tremitates sũt duo pũcta. ¶ Linea recta
ẽ ab vno pũcto ad aliũ breuissima extẽ/
sio i extremitates suas vtrũq3 eoꝝ reci
piens. ¶ Supficies ẽ q̃ lõgitudinẽ ⁊ lati
tudinẽ tñ h3: cuiꝰ termi quidẽ sũt linee.
¶ Supficies plana ẽ ab vna linea ad a/
liã extẽsio i extremitates suas recipiẽs
¶ Angulus planus ẽ duarũ linearũ al/
ternus ꝯtactus: quaꝝ expãsio ẽ sup sup/
ficiẽ applicatioq3 nõ directa. ¶ Quãdo aũt angulum ꝯtinẽt due
linee recte rectilineꝰ angulus noiat̃. ¶ Qñ recta linea sup rectã
steterit duoq3 anguli vtrobiq3 fuerit eq̃les: eoꝝ vterq3 rectꝰ erit
¶ Lineaq3 linee supstãs ei cui supstat ppendicularis vocat̃. ¶ An
gulus vo qui recto maior ẽ obtusus dicit̃. ¶ Angulꝰ vo minor re
cto acutꝰ appellat̃. ¶ Terminꝰ ẽ qd vniuscuiusq3 finis ẽ. ¶ Figura
ẽ q̃ tmino vl termis ꝯtinet̃. ¶ Circulꝰ ẽ figura plana vna qdem li/
nea ꝯtẽta: q̃ circũferentia noiat̃: in cuiꝰ medio pũctꝰ ẽ: a quo oẽs
linee recte ad circũferẽtiã exeũtes sibiinice3 sut equales. Et hic
quidẽ pũctꝰ cẽtrũ circuli dr̃. ¶ Diameter circuli ẽ linea recta que
sup eiꝰ centꝝ trãsiens extremitatesq3 suas circũferẽtie applicans
circulũ i duo media diuidit. ¶ Semicirculus ẽ figura plana dia/
metro circuli ⁊ medietate circũferentie ꝯtenta. ¶ Portio circu/
li ẽ figura plana recta linea ⁊ parte circũferẽtie ꝯtẽta: semicircu/
lo quidẽ aut maior aut minor. ¶ Rectilinee figure sũt q̃ rectis li/
neis cõtinent̃ quarũ quedã trilatere q̃ tribꝰ rectis lineis: quedã
quadrilatere q̃ qtuor rectis lineis. qdã mltilatere que pluribus
q3 quatuor rectis lineis continent̃. ¶ Figurarũ trilaterarũ: alia
est triangulus hñs tria latera equalia. Alia triangulus duo hñs
eq̃lia latera. Alia triangulus triũ inequalium laterũ. Haꝝ iterũ
alia est orthogoniũ: vnũ .s. rectum angulum habens. Alia ẽ am/
bligonium aliquem obtusum angulum habens. Alia est oxigoni
um: in qua tres anguli sunt acuti. ¶ Figurarũ autẽ quadrilateraꝝ
Alia est q̃dratum quod est equilaterũ atq3 rectangulũ. Alia est
tetragonꝰ longꝰ: q̃ est figura rectangula: sed equilatera non est.
Alia est helmuaym: que est equilatera: sed rectangula non est.

De principijs p se notis: ⁊ pmo de diffini/
tionibus earundem.

First page of the first printed edition of Euclid's *Elements*, published in Latin in 1482 factored until we find a *prime* factor. Being a factor, it goes into $n! + 1$. Therefore it is not any of the numbers from 1 to n. So we have found a prime number (either $n! + 1$ or one of its prime factors) greater than n. Since we can do this for any n, the sequence of primes goes on forever.

3.

What goes in must come out

The Sieve of Eratosthenes

Suppose you were asked to make a list of all the prime numbers up to 100. You would start out quickly enough, "1, 2, 3, 5, 7, 11, . . ." and presently would stop to think. How soon you have to stop would depend on how well you know the multiplication table. You might get past the 30's, 40's, and 50's at full speed, but you are not likely to know right away which of the numbers 91 and 97 is prime. And after 100 you would certainly slow down.

One of the first things that would take up time is dividing by 7. If a number is divisible by 5, you can see that immediately, because it will end in 5 or 0. And the even numbers (they are divisible by 2) all end in 2, 4, 6, 8, or 0. So the only numbers that have a chance of being prime, besides 2 and 5 themselves, are those that end in 1, 3, 7, or 9. The multiples of 3 are easily recognized by adding up the digits—a trick you probably know. Thus, 15 is a multiple of 3, because $1+5=6$, which is a multiple of 3. For the same reason 51, 105, 510, and 1005 are multiples of 3. But 67 is not, because $6 + 7 = 13$, which is not a multiple of 3.

If you write down the numbers ending in 1, 3, 7, and 9, leaving out the multiples of 3, you will have mostly prime numbers at first. Presently, though, you will include some numbers which are not prime because they are multiples of 7. Therefore you must divide each candidate by 7 to see whether there is a remainder. After a while you will run into multiples of 11. These stand out, below 100, because the two digits are equal, as in 77 or 88. They also are easy to recognize when they have three digits. The middle digit is either equal to the sum of the other two, as in 253 and 176, or is 11 less than that sum, as in 506 and 847. But soon you will have to divide each possible prime by 13, 17, 19, and so on, and your progress will be much slower.

If your list is to go up to 1000, let us say, there is a more efficient way to make it. Instead of considering one number at a time and checking all possible factors, you can take one factor at a time and find all the numbers up to 1000 that it goes into. Thus you list *all* the numbers from 1 to 1000. Then you cross out all the even numbers, except 2, then all the multiples of 3, except 3, then every 5th number after 5, then every 7th after 7, and so on. (It is obviously sufficient to consider prime factors. For example, you need not cross out multiples of 6 because they are also multiples of 2 and have already been crossed out.) Any number that is not prime will be crossed out sooner or later, and eventually the remaining numbers will be a list of all the primes.

This method is called the *sieve of Eratosthenes.* Whoever named it, must have had a mental picture of a sieve with many layers. Each layer corresponds to a factor and has wire mesh spaced according to the size of the factor. Thus, the 7 layer has wires 7 units apart. Most numbers can slip through the holes, but 14, 21, 28, etc. are stopped by the wires. The numbers are "strained" through one layer after another, and the prime numbers are the ones that go all the way through.

The namer of this method must also, of course, have had a mental picture of Eratosthenes, who invented it. (His name rhymes with "across the knees.") Eratosthenes lived a little later than Euclid. He is most famous for measuring the size of the earth, which is actually quite possible without traveling more than a hundred miles from home. He was fond of gadgets and tricks. Among the Greek thinkers, he gives a particularly modern impression. Anyone who reads nowadays of the work of Greek mathematicians and philosophers begins to feel that they were blind in certain directions—that some ideas which we find natural were impossibly foreign for them. This is because every age has its way of thought. They would probably have found us equally blind in other spots. But a few of the ancient writers give the impression of seeing things just as we do. One of these was Archimedes, the greatest ancient mathematician. Another was Eudoxus, who worked out some of the important ideas in Euclid's *Elements*. A third was Lucretius, who lived two centuries later and wrote in Latin but had studied in the Greek tradition. Eratosthenes also gives the "modern" impression, although his achievements are much less significant than those of Archimedes and Eudoxus.

I have left out one important point in describing the sieve method. How many "layers" must you strain the numbers through before you have only primes left? In other words, how many factors must you test? Suppose the list of primes is to go up to 100. To be sure that 97 is prime, do you have to divide it by every number less than 97? Obviously not, because anything greater than half of 97 can't possibly go in exactly. But as soon as you have passed the first layer—that is, crossed out all the multiples of 2—you know that 97, which is not crossed out, is not exactly double any whole number. Therefore nothing greater than a third of 97 can go into it exactly. If we continue this reasoning, we see that you only have to test factors up to 10, because if 97 has a factor greater than 10, it must have *another* factor less than 10. For example, if 97 could be divided by 13 without remainder (it can't), the quotient would be less than 10 (since 10 thirteens are more than 100) and this quotient would be a factor of 97.

Since only the prime factors have to be tested, the only

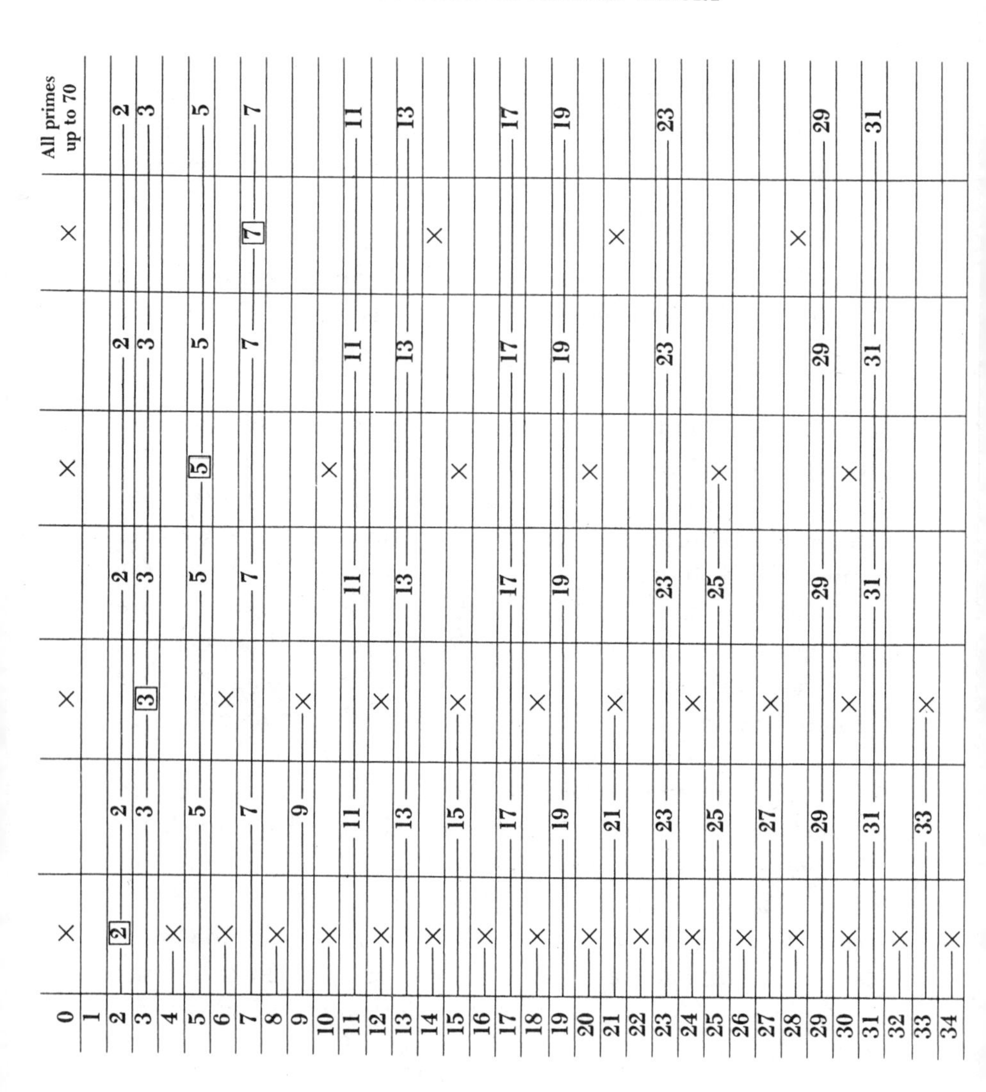

The sieve of Eratosthenes, for numbers up to 70. The ×'s form the sieve, and the numbers pass through from the left. A number is stopped when it meets an ×. The boxed number in each column of ×'s tells how far apart the ×'s are spaced in that column. The boxed number itself is allowed to pass through. A prime number is one that dodges all the ×'s until it reaches the column in which it is boxed.

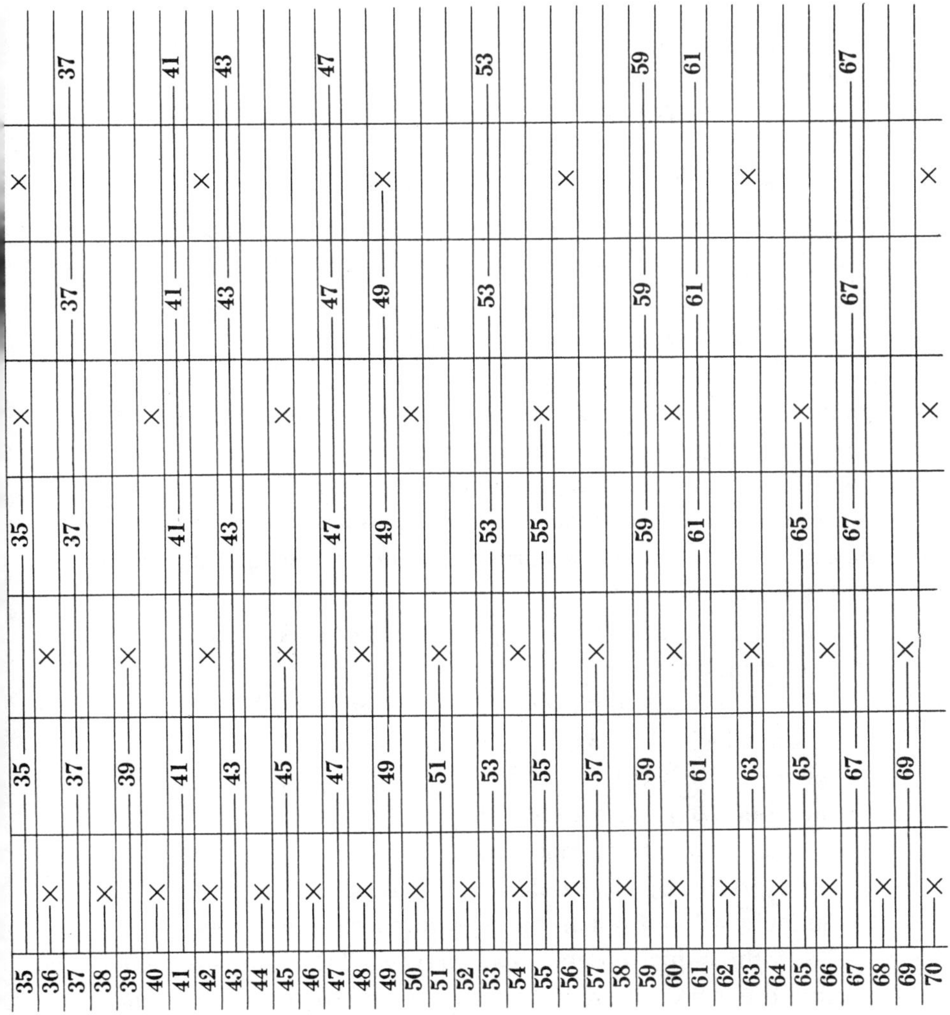

Note that the smallest number eliminated by the 3-spaced column of ×'s is 9, the smallest number eliminated by the 5-spaced column is 25, and so on. Thus the 11-spaced column will not eliminate any number below 121. Since this diagram only goes up to 70, the four columns shown are the only ones needed here. All the numbers that reach the right-hand side are prime.

ones to be considered, for a list of primes up to 100, are 2, 3, 5, and 7. Anything not divisible by one of these is prime. Now 2, 3, and 5 can be tested at a glance. After you cross out the multiples of 2, 3, and 5 up to 100, only three numbers are left that are not prime. These are $49 = 7 \times 7$, $77 = 7 \times 11$, and $91 = 7 \times 13$. Seventy-seven is easily recognized, and 49 is familiar as the square of 7. So 91 is the only number below 100 that looks prime but isn't. If you memorize the number 91, you will be able to tell instantly whether any two-digit number is prime.

To list the primes up to 1000, you must test the prime factors up to 31. This is because 32×32 is more than 1000. Consequently, if a number less than 1000 has a factor at least as big as 32, it must have another factor less than 32. So the procedure for finding whether a three-digit number is prime (I assume now that you are testing a particular number instead of making a list) is as follows.

Factors 2 and 5: If the number does not end in 1, 3, 7, or 9, it is not prime.

Factor 3: If the digits add up to a multiple of 3, the number is not prime.

Factor 11: If the outer two digits add up to the inner one, the number is not prime. Or if they add up to 11 more than the inner one, the number is not prime.

Factors 7, 13, 17, 19, 23, 29, 31: If any of these goes into the number without remainder, the number is not prime.

Any three-digit number that survives all these tests is prime. Most of the work is in dividing by 7, 13, and so on. But you don't have to finish each division; you can leave it undone as soon as you see that there is a remainder. You don't usually have to go all the way to 31, either. You can stop as soon as the prime factor you are about to test has a square that is larger than the number you are working on. If the number is less than 500, you can stop after 19. If it is less than 250, you can stop after 13.

Even if the number is very large (but less than 1000) you can do without 29 and 31. Just remember the three numbers $841 = 29 \times 29$, $899 = 29 \times 31$, and $961 = 31 \times 31$. If a three-digit number is not one of these three, and has no factor up to 23, it is prime. With practice, a number can be tested in a minute or two.

You can see, now, why I said in Chapter 2 that it is only a day's work to find out that 65537 is prime. Since $65537 = 256^2 + 1$, you need only check all the prime factors up to 256. There are about fifty of these. Suppose each one takes five minutes—it is a matter of dividing 65537 by a two- or three-digit number. The whole problem would take about four hours if you worked without stopping. If you rest a lot, you can still do it in a day. But there is a quicker way (see Chapter 6).

The Chemistry of Numbers

The prime numbers are like building blocks. By multiplying them together, you can make any number you like. Or you can take a number and break it down into prime factors. Thus 60 is twice 30, 30 is twice 15, and $15 = 3 \times 5$. So $60 = 2 \times 2 \times 3 \times 5$. Or you may start out by writing $60 = 10 \times 6$. Then $10 = 2 \times 5$ and $6 = 2 \times 3$. So $60 = 2 \times 5 \times 2 \times 3$. Again, $60 = 20 \times 3$, $20 = 5 \times 4$, $4 = 2 \times 2$. This gives $60 = 5 \times 2 \times 2 \times 3$.

No matter how you break 60 down, you will alwavs get a 5, a 3, and two 2's, in some order. You will never get a factor 7, or *two* 3's, or only one 2, or be missing 5 altogether. This is an example of the *unique factorization theorem:* Any number can be obtained in just one way by multiplying together prime factors. (It is understood that the order of the factors doesn't matter. That is, 2×3 and 3×2 are the same way of getting 6.

Chemists of the 16th century, engaged in the process of "distillation," from an engraving by the Flemish artist Johannes Stradanus (1523–1605). The artist's caption reads: "In the fire,

the juice of all bodies is turned by art into a mighty billow, clear and most potent."

But the number of times each factor occurs does matter. In 60, 2 always occurs twice, and 3 and 5 once each.)

The unique factorization theorem is so important that it is also called the *fundamental theorem of arithmetic.* Until you have mastered it, you cannot really understand the chemistry of numbers; you are like the alchemists of medieval times who studied ways of transforming one material into another, just as modern chemists do. They learned how to make many interesting substances, but they did not understand the laws governing chemical elements, and consequently they were forever attempting impossible tasks, such as making gold out of rocks.

What we call modern chemistry began around the year 1800, when two things became clear. First, there are certain substances (elements) out of which everything else is made. An element can never be made from a combination of other elements, or broken down into other elements. The transformations that substances undergo—burning, baking, boiling, freezing, dissolving, settling—are all due to the arrangement of elements into different combinations. No matter what you do to your caldron (as long as nothing escapes or leaks in, even gas) it will contain the same elements that were in it originally, and no others. And second, every element is present in the exact amount that you started with.* What goes in must come out. Or, you may put it, nothing ever comes out that didn't go in.

The alchemists had some glimmering of this, but they had no true idea of which substances are elements. Ancient writers had suggested fire, air, earth, and water; many alchemists believed in salt, sulphur, and mercury. In reality there are over ninety different elements, and some of the most important ones are odorless, colorless gases which the alchemists didn't know from air. They also had no idea of the second part of the law, that the exact amount of each element would remain constant. In fact, this seemed untrue, because a whole caldron could boil off, leaving

* See note in Appendix 1.

only a little sediment. They didn't realize that escaping gas was important, and anyway their weighing scales weren't very good. The advances in chemistry around 1800 took place because chemists were beginning to use very accurate scales and were keeping track of gases.

In number theory, the prime numbers are like elements. If you multiply any group of numbers together and break the result down into prime factors, you will have just the prime factors of the numbers you started with, and each prime factor will occur just as many times as it did in all the original numbers combined. For example, multiply $2 \times 28 \times 21$. The prime factor 2 occurs once in 2 and twice in 28—three times altogether. The factor 3 occurs just once, in 21. And 7 occurs twice, once in 28 and once in 21. So three 2's, two 7's, and a 3 have gone in. What comes out?

$$\begin{aligned} 2 \times 28 \times 21 &= 1176 \\ &= 8 \times 147 \\ &= 8 \times 3 \times 49 \\ &= 2^3 \times 3 \times 7^2 \end{aligned}$$

as expected.

You cannot tell at a glance that 1176 has a factor 7. For example, the presence of the digit 7 is no clue. The number 1716 has the same digits but is not divisible by 7. This is a mistake often made by the alchemists. They supposed that you could make gold by mixing together the right combination of ingredients having the various qualities of gold, for example, something metallic, something yellow, and a pinch of something shiny. But since gold is actually an element, this could never work unless you included an ingredient which already contained gold, such as gold chloride. And gold chloride is not yellow, or metallic, or shiny.*

The alchemist in number theory might try to make a multiple

* See note in Appendix 1.

of 7 by multiplying together several numbers containing 7 as a digit, thus: $71 \times 76 \times 27 \times 37$. For good measure, since 7 is 1 less than a power of 2, he might multiply by 31×15, since each of these is 1 less than a power of 2. Then, to clinch it, he might raise the result to the 7th power. But if you understand the fundamental theorem of arithmetic, you know without any work that $(71 \times 76 \times 27 \times 37 \times 31 \times 15)^7$ is not a multiple of 7, because none of the numbers 71, 76, 27, 37, 31, 15 is a multiple of 7. What goes in must come out.

The Greatest Common Factor

Proving the unique factorization theorem is not easy. It is obvious that when two numbers are multiplied together their factors will also be factors of the product. The hard part is to show that the product will have no *other* prime factors. For example, if two numbers which are not multiples of 7 are multiplied together, the result cannot be a multiple of 7. Try this a few times to see that it is true. It does not work if 7 is replaced by a number that is not prime. Thus, if you multiply 15 by 8, you get 120, which is a multiple of 6 even though 15 and 8 are not. The reason this can happen is that 15 contains the factor 3, and 8 contains 2. When you put them together, the 3 and 2 can combine to make 6. The same thing happens in chemistry. Salt is not an element; it contains the elements sodium and chlorine. Therefore it can be made by mixing bicarbonate of soda (which contains sodium) with hydrochloric acid (which contains chlorine) even though neither contains whole salt.

To prove the unique factorization theorem, we must proceed in three steps. First we must prove an important theorem called the *greatest common factor* theorem or GCF theorem. This is the heart of the matter. Second, using the GCF theorem, it will be easy to prove the statement in the last paragraph, that if p is prime and ab is a multiple of p, then either a or b must be a multiple of p. And finally, once this is known, the unique factorization theorem will follow without difficulty.

The GCF theorem has to do with a factor that goes exactly into *each* of two given numbers. Thus 4 is a *common factor* of 12 and 16, but 3 is not, since it goes into 12 but not into 16. 2 and 1 are also common factors of 12 and 16, but 8 is not, since it doesn't go into 12.

The GCF theorem says that if m and n are any two numbers, it is possible to choose two other numbers c and d so that $cm + dn$ is a common factor of m and n. For example, if $m = 26$ and $n = 18$, we choose $c = -2$, $d = 3$, and then

$$\begin{aligned} cm + dn &= -2 \times 26 + 3 \times 18 \\ &= -52 + 54 = 2 \end{aligned}$$

which is a common factor of 26 and 18. Note that c and d are allowed to be negative or zero, but not fractions.

To prove that this can always be done, let us use the method of infinite descent already mentioned in Chapter 2. Suppose that we have two numbers m and n that *don't* obey the GCF theorem. (That is, there are no suitable numbers c, d.) Call the larger one m.

Now, m cannot be divisible by n, for then n would be a common factor of m and n, and we could choose $c = 0$, $d = 1$. Therefore if m is divided by n there will be a remainder r which is less than n. Thus

$$m = nq + r$$

where q is the quotient.

But it follows that the pair of numbers n, r also don't obey the GCF theorem. For if they did, they would have a common factor

$$\begin{aligned} f &= sn + tr \\ &= sn + t(m - nq) \\ &= sn + tm - tnq \\ &= tm + sn - tqn \\ &= tm + (s - tq)n \\ &= cm + dn \end{aligned}$$

where we set $c = t$, $d = s - tq$. And f would be a factor not only of n and r but also of m, which is r plus a multiple of n. Hence m and n would obey the GCF theorem, which we assumed is not so.

Dividing n by r, we obtain a new remainder r', less than r, and by the same reasoning r and r' don't obey the theorem. Thus we can construct an infinite downward sequence of positive numbers, m, n, r, r', and so on. This is impossible, and so m and n must obey the theorem after all.

Actually, the downward sequence is not infinite but stops when it reaches the GCF itself. Thus, let $m = 26$, $n = 18$. Dividing 26 by 18, we find a remainder 8. Dividing 18 by 8, we find a remainder 2. Dividing 8 by 2, we find no remainder, which means that 2 is the GCF. Since it goes into 8, it must go into $18 = 8 + 8 + 2$, and also into $26 = 18 + 8$. To find c and d, we note that

$$8 = 26 - 18$$

and

$$\begin{aligned} 2 &= 18 - 8 - 8 \\ &= 18 - (26 - 18) - (26 - 18) \\ &= -2 \times 26 + 3 \times 18 \end{aligned}$$

therefore $c = -2$, $d = 3$.

The name greatest common factor comes from the fact that *any* common factor of m and n must also be a factor of $cm + dn$. (For example, if it goes twice into m and five times into n, then it will go $2c + 5d$ times into $cm + dn$.) Therefore the common factor, $f = cm + dn$, selected by the GCF theorem is actually the greatest of all the common factors of m and n, being a multiple of all the others. It is also called the greatest common divisor, since "divisor" means the same as "factor."

Now for the second step. Consider a prime number p, and suppose that ab is a multiple of p. Let f be the greatest common factor of a and p. Then because of the GCF theorem, we can write

$$f = cp + da$$

Since p is prime, there are only two possibilities: either $f = p$, or $f = 1$. But if $f = p$, then p is a factor of a. And if $f = 1$, then

$$\begin{aligned} b = bf &= b(cp + da) \\ &= (bc)p + d(ab) \end{aligned}$$

which shows that p is a factor of b (remember that p goes into ab). Therefore one of the two numbers, a and b, is divisible by p.

The third step completes the proof. Suppose that a certain number could be factored in two ways

$$N = p_1p_2p_3 \ldots p_m = q_1q_2q_3 \ldots q_n$$

Here all the p's and q's are prime, and some of them may be equal to each other. Unique factorization says merely that any prime number that appears among the p's must appear the same number of times among the q's. If this were not so, we could divide both sides of the equation by this prime as often as necessary until it appeared no longer on one side (say among the q's) but still appeared on the other (among the p's). We should then have a product

$$r_1r_2 \ldots r_k$$

where the r's are simply those q's that remain after the division, none of which are equal to the prime p by which we have divided. But the whole product is divisible by p since it is equal to the left side in which p still appears. This is impossible. Each of the numbers r_1, r_2, etc. is prime and different from p; therefore none of them is divisible by p. It follows that r_1r_2 cannot be divisible by p. Therefore $r_1r_2r_3$ cannot be divisible by p, and so on. Thus $r_1 \ldots r_k$ is not divisible by p, and the unique factorization theorem cannot be false.

The Aliquot Giver

Now that we have progressed beyond numerical alchemy, we know enough to tackle one of the favorite subjects of Pythagoras—perfect numbers. A number is perfect if it is the sum of all its factors, like the number 6. In this game you are supposed to count 1 as a factor, but not to count 6 itself. Thus, the factors of 6 are 1, 2, and 3, which add up to 6. Another perfect number is 28, which has factors 1, 2, 4, 7, 14. Most numbers are either excessive, like 14, or defective, like 20 (14 exceeds the sum of its factors, whereas 20 is less than the sum of its factors).

When Pythagoras said perfect, he meant perfect. Why was he so impressed by a number that is equal to the sum of its factors? I have a hunch about this, connected with the way in which fractions were written in his time.

The Egyptians, from whom Pythagoras studied, had no short way of writing a fraction like ⅔ or ⅜. The only fractions for which they had a simple symbol were the aliquot fractions. An aliquot fraction has a numerator 1, like ⅓ or ⅛. They wrote these fractions much as we do in words: "the third part" or "the eighth part."

This left them in a difficult position when it came to adding fractions. If we are asked to add $\frac{1}{3}$ to $\frac{1}{8}$, we reduce both fractions to a common denominator and say $\frac{8}{24} + \frac{3}{24} = \frac{11}{24}$. But the Egyptians had no symbol for $\frac{11}{24}$. It did not occur to them to write "eleven of the twenty-fourth part"—perhaps this seemed roundabout and ugly. To them it seemed easier and more direct to write "an eighth part and a third part." So it would be meaningless to ask them, "How much is $\frac{1}{8} + \frac{1}{3}$?" It is as if someone asked us "How much is two hundred and forty-seven?"

However, they did understand the idea of the common denominator. A typical problem for them was this: A farmer had 24 sheep. A third died of starvation and an eighth were carried away by wolves. How many sheep had the farmer lost? The answer would be 11.

This problem could be reversed, as follows: A farmer lost 11 out of 24 sheep. What fraction had he lost? In this form the problem is a trivial one for us, since the answer is $\frac{11}{24}$. But for the Egyptians it was a difficult problem, since they could not write $\frac{11}{24}$. To "find the fraction" that the farmer had lost meant finding aliquot fractions that add up to $\frac{11}{24}$. It was their custom not to use the same aliquot fraction more than once in a sum; thus, they would not write

$$\frac{1}{24} + \frac{1}{24} + \frac{1}{24} + \frac{1}{24} + \frac{1}{24} + \frac{1}{24} + \frac{1}{24} + \frac{1}{24} + \frac{1}{24} + \frac{1}{24} + \frac{1}{24}$$

The *answer* to the problem, in their language, was $\frac{1}{3} + \frac{1}{8}$.

Here are some more problems of the same kind. (The answer is required to be a sum of aliquot fractions with no repetitions.)

What fraction is 5 of 6? (Answer: $\frac{1}{2} + \frac{1}{3}$.)

What fraction is 8 of 20? (Answer: $\frac{1}{4} + \frac{1}{10} + \frac{1}{20}$. We can't use $\frac{1}{5}$ and $\frac{1}{5}$ since each aliquot can be used only once.)

What fraction is 5 of 12? (Two answers: $\frac{1}{3} + \frac{1}{12}$ or $\frac{1}{4} + \frac{1}{6}$.)

What fraction is 17 of 24? (Five answers.)

What fraction is 7 of 10?

What fraction is 3 of 14?

What fraction is 13 of 12? (Here you are not allowed to say "$1/1 + 1/12$." A whole is not to be used. A correct answer is $1/2 + 1/3 + 1/4$.)

What fraction is 43 of 40? (Do not use a whole.)

All the problems above have answers in which the aliquot parts have the common denominator given in the problem. Thus, in the first problem, $1/2$ and $1/3$ have the common denominator 6. But it is not always possible to find an answer with the given denominator as common denominator. For example, what fraction is 2 of 3?

The simplest answer, in aliquot fractions, is that $2/3 = 1/2 + 1/6$. But $1/2$ and $1/6$ do not have the common denominator 3. If a farmer had 3 sheep and lost 2 of them, it is not possible that $1/2$ of his sheep died of cholera and $1/6$ were carried off by wolves (unless the wolves carried off a sheep already *half*-dead of cholera—a trick answer not allowed). But if he started with 6 sheep, he might have lost 3 sheep ($1/2$) to cholera and 1 sheep ($1/6$) to wolves —4 sheep in all. Thus 4 out of 6 can be written in aliquots, whereas 2 out of 3 cannot.

This suggests a new problem: what is the largest fraction that can be written in aliquots with a given common denominator? This can be easily solved, by adding up all the aliquot parts of the given denominator. If the denominator is 8, the aliquot parts are 1, 2, and 4. So the largest fraction that can be written in aliquot parts of 8 is $7/8 = 1/8 + 2/8 + 4/8 = 1/8 + 1/4 + 1/2$.

What is the largest fraction that can be written in aliquot parts of 14? (Answer: $10/14 = 1/2 + 1/7 + 1/14$.)

What is the largest fraction that can be written in aliquot parts of 63? (Answer: $41/63$.)

What is the largest fraction that can be written in aliquot parts of 20? (Answer: $1/2 + 1/4 + 1/5 + 1/10 + 1/20 = 22/20 = 1\,2/20$.)

These problems can be put into practical form. Suppose you have 14 books and decide to give a different aliquot fraction of them to each of your friends. Then you will give 7 books (½) to Lucy, let us say, 2 books (1/7) to Linus, and 1 book (1/14) to Charlie Brown. At this point you cannot give any more, since you have used all the aliquot fractions. But you still have 4 books that you have no use for, since you have read them all.

On the other hand, let us say, that you set out to give away 20 pencils. After giving 10 to Lucy (½), 5 to Linus (¼), and 4 to Schroder (⅕), you have only 1 left (1/20), which you give to Snoopy. Now comes Charlie Brown and finds you empty-handed, although you have promised him 2 pencils (1/10).

It seems that "aliquot gifts" are a poor idea, since you never have the right number to give away. But the secret is in knowing what number to start with. The wise "aliquot giver" will give away 6 candy bars at a time. He will give 3 to Lucy, 2 to Schroder, and 1 to Charlie Brown (½, ⅓, ⅙). Thus, he exhausts the aliquot fractions and has the right number to go around. For the aliquot giver, 6 is a perfect number, whereas 14 is excessive (you have some left over) and 20 is defective (you haven't enough to go around).

In terms of fractions, the special characteristic of 6 is that the largest fraction which can be written in aliquot parts of 6 is exactly $1 = \frac{1}{2} + \frac{1}{3} + \frac{1}{6}$. But the same idea can be expressed without fractions, in terms of factors. When the aliquot giver gives away 14 books, he gives each factor of 14 to a different friend. The number of books he gives away is just the sum of the factors of 14, $7 + 2 + 1 = 10$. From this point of view, the merit of 6 is that it equals the sum of its factors. Thus, the numbers that Pythagoras called excessive, defective, or perfect are just those that are excessive, defective, and perfect for aliquot giving.

Adding Up Factors

Once you have chosen aliquot giving as a way of life, you will want to know as many perfect numbers as possible. You can give away 6 candy bars at a time, and 28 books, but if you are distributing popcorn or (shall we say) peanuts, you will want some perfect number in the hundreds. How can other perfect numbers be found?

This is where the unique factorization theorem comes in. Take a number and break it into prime factors, thus

$$60 = 2^2 \times 3^1 \times 5^1 \text{ or}$$
$$40 = 2^3 \times 5^1 \text{ or}$$
$$36 = 2^2 \times 3^2$$

If the number is called n, then we can write

$$n = p^a q^b r^c \ldots$$

where p, q, r, etc., are the prime factors and a, b, c, etc., are the powers to which they must be raised to give a product equal to n.

Now suppose f is a factor of n. Then any prime factor of f is a factor of n, so that f can have no prime factors besides p, q, r, etc. Thus

$$f = p^x q^y r^z \ldots$$

where x, y, z, etc. are a new set of exponents. To be sure, f may not contain all the prime factors of n. But then it can still be written as above, provided that some of the exponents are zero. The meaning of p^0 is 1. Thus, if f is not divisible by p, we can still write

$$f = p^0 q^y r^z \ldots = q^y r^z \ldots$$

The exponent zero simply means that the factor p occurs no times.

On the other hand, the exponent x cannot be bigger than a, for if p^{a+1} were a factor of f, it would also be a factor of n. In that case p would have to be a factor of $q^b r^c \ldots$, which is impossible because of the unique factorization theorem. Therefore x can be anything from 0 to a, y can be anything from 0 to b, and so on.

We can now list all the factors of n and find their sum. First, let all the exponents be zero. Then

$$f = p^0 q^0 r^0 \ldots = 1$$

Now let x take the values 1, 2, etc., up to a, while the other exponents are still zero. Then

$$\begin{array}{ll} \text{if } x = 1, & f = p \\ \text{if } x = 2, & f = p^2 \\ & \\ \text{if } x = a, & f = p^a \end{array}$$

Adding these together, we find the sum

$$1 + p + p^2 + \cdots + p^a$$

Next, let $y = 1$. Then x can still be anything from 0 to a. Thus

$$\begin{array}{ll} \text{if } x = 0, & f = q \\ \text{if } x = 1, & f = pq \\ \text{if } x = 2, & f = p^2 q \\ & \\ \text{if } x = a, & f = p^a q \end{array}$$

Adding this second group together, we get

$$q + pq + p^2 q + \cdots + p^a q \text{ or } (1 + p + p^2 + \cdots + p^a) \times q$$

By letting $y = 2$, we get a third group of factors, which add up to $(1 + p + p^2 + \cdots + p^a) \times q^2$. We can go on until $y = b$, which gives a group adding up to $(1 + p + p^2 + \cdots + p^a) \times q^b$. If we combine all these groups into a super-group and add them together, the sum is obviously

$$(1 + p + p^2 + \cdots + p^a) \times (1 + q + q^2 + \cdots + q^b)$$

This super-group does not contain all the factors of n, if n has a third prime factor r. By letting $z = 1$, we get a second super-group. The factors in the second super-group add up to

$$(1 + p + p^2 + \cdots + p^a) \times (1 + q + q^2 + \cdots + q^b) \times r$$

By letting z take all values up to c, we get several super-groups, which may be combined into a grand super-group with a sum of

$$(1 + p + p^2 + \cdots + p^a) \times (1 + q + q^2 + \cdots + q^b) \times (1 + r + r^2 + \cdots + r^c)$$

If n has a fourth prime factor s, then each power of s will correspond to another grand super-group. By this time it is obvious that the sum of *all* the factors of n is

$$(1 + p + \cdots + p^a) \times (1 + q + \cdots + q^b) \times (1 + r + \cdots + r^c) \times \ldots$$

where the product continues as long as there are more prime factors. However, we must take note that the number n is itself included as a factor, according to this formula. Our method of generating factors allows the possibility that all the exponents have their maximum values: $x = a$, $y = b$, etc. In this case $f = n$.

Let us try this formula on a few numbers.

(1) $n = 9 = 3^2$. Then $p = 3$, $a = 2$, and q, r, etc. do not exist. The formula is then

$$1 + p + p^2 = 1 + 3 + 3^2 = 1 + 3 + 9 = 13$$

This is obviously the sum of the factors of 9, including 9 itself.

(2) $n = 15 = 3 \times 5$. Then $p = 3$, $a = 1$; $q = 5$, $b = 1$; and r, s, etc. do not exist. In this case the formula is

$$(1 + p) \times (1 + q) = (1 + 3) \times (1 + 5) = 4 \times 6 = 24$$

The factors of 15, including 15 itself, are 1, 3, 5, and 15. They do add up to 24.

(3) $n = 126 = 2 \times 3^2 \times 7$. Then $p = 2$, $a = 1$; $q = 3$, $b = 2$; $r = 7$, $c = 1$. Here the formula gives

$$(1 + p)(1 + q + q^2)(1 + r) = 3 \times 13 \times 8 = 39 \times 8 = 312$$

Adding the factors directly, we find

$$1 + 2 + 3 + 6 + 9 + 18 + 7 + 14 + 21 + 42 + 63 + 126 = 312$$

Clearly, it is easier to use the formula than to add up all the factors. The formula is also more reliable, because you might easily leave out a factor or make a mistake in a long addition.

The Perfect-Number Equation

A perfect number is the sum of all its factors, *not* counting the number itself. Therefore, if this formula is applied to a perfect number, it will give twice the number. Thus

$$n = 6 = 2 \times 3. \text{ Then } p = 2,\ a = 1;\ q = 3,\ b = 1.$$
$$(1 + p)(1 + q) = 3 \times 4 = 12 = 2n$$

$$n = 28 = 2^2 \times 7. \text{ Then } p = 2,\ a = 2;\ q = 7,\ b = 1.$$
$$(1 + p + p^2)(1 + q) = 7 \times 8 = 56 = 2n$$

So, in order to find a perfect number, we must find a set of primes $p, q, r, \ldots$ and exponents $a, b, c, \ldots$ such that

$$(1 + p + \cdots + p^a)(1 + q + \cdots + q^b)(1 + r + \cdots + r^c) \ldots = 2p^a q^b r^c \ldots$$

Now, if one of the prime factors is 2, we can draw some interesting conclusions. Suppose that $p = 2$. Then

$$1 + p + \cdots + p^a = 1 + 2 + 4 + \cdots + 2^a = 2^{a+1} - 1$$

(This was proved in Chapter 2. For example, $1 + 2 = 2^2 - 1$; $1 + 2 + 2^2 = 2^3 - 1$.) In this case, the perfect-number equation becomes

$$(2^{a+1} - 1)(1 + q + \cdots + q^b)(1 + r + \cdots + r^c) \dots$$
$$= 2 \times 2^a q^b r^c \cdots = 2^{a+1} q^b r^c \dots$$

This shows that $(2^{a+1} - 1)$ is a factor of $2^{a+1} q^b r^c \dots$. Hence every prime factor of $2^{a+1} - 1$ is one of the numbers 2, q, r, etc. But 2 obviously is not a factor of $2^{a+1} - 1$. Furthermore, we assumed from the start that a is not zero, which makes $2^{a+1} - 1$ greater than 1. (This is equivalent to assuming that the perfect number we are dealing with is even.) Therefore $2^{a+1} - 1$ has at least one prime factor greater than 1, which must be one of the primes q, r, etc. We may just as well suppose that it is q. Then

$$2^{a+1} - 1 = qk$$

where k is some whole number. Now the perfect-number equation can be written

$$qk(1 + q + \cdots + q^b)(1 + r + \cdots + r^c) \dots$$
$$= (qk + 1) q^b r^c \dots$$

It is revealing to express this in fractions. Dividing both sides by $qk \times q^b r^c \dots$, we get

$$\frac{(1 + q + \cdots + q^b)(1 + r + \cdots + r^c) \cdots}{q^b r^c \dots} = \frac{qk + 1}{qk}$$

Now

$$\frac{1 + q + \cdots + q^b}{q^b} \text{ is the same as } 1 + \frac{1}{q} + \cdots + \frac{1}{q^b}$$

Therefore

$$\left(1+\frac{1}{q}+\cdots+\frac{1}{q^b}\right)\left(1+\frac{1}{r}+\cdots+\frac{1}{r^c}\right)\ldots=1+\frac{1}{qk}$$

In this last equation, the quantity on the right side cannot be bigger than $1+\frac{1}{q}$, since $\frac{1}{qk}$ cannot be bigger than $\frac{1}{q}$. But the quantity on the left side cannot be *smaller* than $1+\frac{1}{q}$. This is because we have assumed that $2^{a+1}-1$ is divisible by q. Therefore b is at least 1, and the quantity $\left(1+\frac{1}{q}+\cdots+\frac{1}{q^b}\right)$ must be at least as big as $1+\frac{1}{q}$. The other factors $\left(1+\frac{1}{r}+\cdots+\frac{1}{r^c}\right)$, etc. cannot be less than 1. Therefore the product

$$\left(1+\frac{1}{q}+\cdots+\frac{1}{q^b}\right)\left(1+\frac{1}{r}+\cdots+\frac{1}{r^c}\right)\ldots$$

is no less than $1+\frac{1}{q}$.

It follows that both sides of the equation are exactly equal to $1+\frac{1}{q}$. This has three consequences:

$$k=1\left(\text{so that } 1+\frac{1}{qk}=1+\frac{1}{q}\right)$$
$$b=1\left(\text{so that } 1+\frac{1}{q}+\cdots+\frac{1}{q^b}=1+\frac{1}{q}\right)$$

And the factors r, s, etc. do not exist. (Otherwise the factors $\left(1+\frac{1}{r}+\cdots+\frac{1}{r^c}\right)$, etc. would spoil the equation.)

Therefore $q=2^{a+1}-1$, and the perfect number is 2^aq. So we have proved a theorem: every even perfect number is a product of the form $2^a\times(2^{a+1}-1)$, where $2^{a+1}-1$ is a prime number. But if $2^{a+1} - 1$ is not prime, then $2^a \times (2^{a+1} - 1)$ is not perfect. Thus, if $a = 1$, $2^{a+1} - 1 = 3$, and $2^a \times (2^{a+1} - 1) = 6$. If $a = 2$, $2^{a+1} - 1 = 7$, and $2^a \times (2^{a+1} - 1) = 28$. If $a = 3$, $2^{a+1} - 1 = 15$, which is not prime. Therefore $2^a \times (2^{a+1} - 1)$, which is 120, is not perfect.

If $a = 4$, then $2^{a+1} - 1 = 31$, which is prime. This gives us a new perfect number, $16 \times 31 = 496$. Its factors are 1, 2, 4, 8, 16 (the first group, adding up to 31) and 31, 62, 124, 248. It is easy to see that the sum is 496.

If $a = 5$, no perfect number results, since 63 is not prime. But 127 is prime, and therefore $64 \times 127 = 8128$ is perfect. The aliquot giver can give away 496 crackerjacks at a time, and 8128 peanuts.

If there is anything of which you have 100,000 to a million or so, you will find it very inconvenient to give it away by the aliquot method. After $a = 6$, the lowest number for which $2^{a+1} - 1$ is prime is $a = 12$. This gives a perfect number, 4096×8191 or 33,550,336. It is not easy to find a use for this number. I suggest that you invest wisely and manage your affairs so that at the end of your life you have exactly $33,550,336. Then you can make an aliquot will, leaving $16,775,168 to the Ford Foundation, $8,387,584 for heart research, and so on, ending with $1 for Charlie Brown.

The formula $2^a \times (2^{a+1} - 1)$ is given in Euclid's *Elements*, along with a proof that it is perfect if $2^{a+1} - 1$ is prime. Euclid did not prove what we have shown just now, that there are no even perfect numbers besides the ones given by this formula. This was first done by Euler, whose name is pronounced "oiler."

Leonhard Euler was Swiss and lived in the eighteenth century. This was the period we call the Age of Enlightenment, when thinking men had great faith in democratic ideas and the scientific, rational approach to all problems. One source of this optimism was the great progress being made in mathematical physics. Late in the seventeenth century, Isaac Newton had developed ways of finding equations to describe nature. Euler was one of the leaders in applying these methods, and found equations that worked very well when applied to such complicated things as the spinning of a top, the vibrations of a string, and the flow of water. He lived a long life and did an enormous amount of work. In the theory of numbers, much of his work was devoted to proving or disproving things that Fermat had said a century earlier. It was Euler, for example, who proved that $65536^2 + 1$ is not a prime number (see Chapters 2 and 6).

Euler's theorem on perfect numbers applies only to even perfects. If n is odd, then $2^{a+1} - 1$ equals 1 and does not contain a factor q; hence the reasoning breaks down. There may, consequently, be odd perfect numbers that are not of the form $2^a(2^{a+1} - 1)$. However, none is known. Whether any odd perfects exist is an open question. It is more difficult to analyze the possibility of odd perfects, because if p is not 2 there is not the convenient relation $1 + 2 \cdots + 2^a = 2^{a+1} - 1$. Instead

$$1 + p + \cdots + p^a = \frac{p^{a+1} - 1}{p - 1}$$

which is not as useful.

I cannot end this chapter without mentioning *amicable numbers.* These are two numbers, such as 220 and 284, each of which is the sum of the *other's* factors. Pythagoras considered such a pair the very essence of friendship. They can be used by aliquot givers in the following way. Your investments have failed, and you have only $284. You make an aliquot will, leaving $142 to your wife, $71 to your daughter, $4 to your barber, $2 to a distant nephew, and $1 to the college of your choice. Since 284 is an excessive number, there is a surplus of $64.

Your friend has fared even worse and reaches his deathbed with only \$220. His aliquot will starts with a bequest of \$110 and continues down to \$1. The total is \$284, which will leave *his* estate \$64 short. Of course the solution is to swap estates. After making your wills, you simply hand over your wallet, with \$284, in exchange for his, with \$220. Then each will can be executed perfectly as written.

If you factor 220 and 284, you will find that $220 = 4 \times 5 \times 11$ and $284 = 4 \times 71$. The secret of their friendship is that $(11 + 1) \times (5 + 1) = (71 + 1)$, and also $(11 + 1) + (5 + 1) = \frac{1}{4}(71 + 1)$. You can easily show that if $xy = 2^a(x + y)$, and $(x - 1)$, $(y - 1)$, and $(xy - 1)$ are all prime, then $2^a(xy - 1)$ and $2^a(x - 1)(y - 1)$ are amicable numbers. Simply use the formula for the sum of factors. Remember to subtract the original number. (In this case, $x = 12$, $y = 6$, $a = 2$.)

By using the unique factorization theorem, you can show (but not easily) that xy can equal $2^a(x + y)$ only if there is a number b such that $x = (2^b + 1)2^a$, $y = (2^b + 1)2^{a-b}$. (I assume that $x > y$. Note that x cannot $= y$, for then $xy - 1$ would not be prime.) Therefore, if a and b can be found so that the three numbers $(2^b + 1)2^a - 1$, $(2^b + 1)2^{a-b} - 1$, and $(2^b + 1)^2 2^{2a-b} - 1$ are prime, then the two numbers $[(2^b + 1)^2 2^{2a-b} - 1]2^a$ and $[(2^b + 1)2^{a-b} - 1]\,[(2^b + 1)2^a - 1]2^a$ are amicable. From $a = 2$, $b = 1$, you get 220 and 284. But other values of a and b give much higher numbers. Thus if $a = 4$, $b = 1$, we find $x - 1 = 47$, $y - 1 = 23$, $xy - 1 = 1151$, all prime. The amicable numbers are $16 \times 1151 = 18416$ and $16 \times 23 \times 47 = 17296$. (Why can't we use $a = 3$, $b = 1$?)

4.

Arithmetica

The Middle Ages

It is a mistake to think that nothing happened during the Middle Ages. For one thing, a lot was going on outside Europe. In the Arab countries a whole religion was born. Persia had caliphs and courtiers, cities and citadels, poets, warriors, and men of wisdom. The tales of Arabian nights come from this time. Greek civilization continued in Egypt and Asia Minor. Twice in a thousand years the whole of Central Asia was swept by wandering tribes as by a wind from the East. And there was China, which has its own history.

Even in Europe things did not stand still. The Roman Empire fell apart by degrees. Then a new empire grew up under Charlemagne and it also fell apart, so that the people remembered it only in legends, just as the Greeks in Homer's time had had legends about the great Mycenaean kings before them.

Two and a half centuries after Charlemagne, the Normans landed at Hastings, and the English language was born. Soon after that, on the Continent, the great cathedrals were begun, monuments of skill and devotion which still decorate the countryside. Those who laid the cornerstone of these cathedrals were dead, and their children too, long before the towers were finished. We could not make such things today, even if we had the arts, for we lack the patience.

The medieval period seems backward in comparison with ancient times. It had no thinkers to match Aristotle, no generals like Alexander, no engineers like those of Rome. Its culture was less brilliant, its ambitions less glorious, its people less literate, its institutions less enlightened. But this comparison is unfair. Greece and Rome, in their days of greatness, were not part of Europe. They were on the edge of a string of nations that encircled the eastern half of the Mediterranean, nations that had already nurtured civilization for as many centuries as have passed since the time of Caesar. The Middle Ages were the beginning of civilization in Europe.

The peoples of Europe—Norsemen, Goths, Franks—had been wanderers a short time before, when the Romans taught them how to live an orderly life and stay in one place. The early Middle Ages were for them a period of activity and development, a rise from barbarism rather than a fall from glory. They made no epics about the Romans; their first romances came from the time of Charlemagne. The Romans had been foreigners. Charlemagne was the first European emperor, and his paladins were the first European heros. It is from them, and from the knights of Arthur of Britain, that we derive the idea of a wandering hero-at-large, like Batman or the Lone Ranger, keeping his eyes peeled for the day's good deed. Robin Hood and Santa Claus, Goldilocks and Cinderella all come to us from the Middle Ages.

But the greatest medieval idea was Christianity. First in the wake of the Roman generals, and again on the banners of the paladins, it was carried all over the continent. To this day no emperor has conquered Europe for long enough to hand it to his successor, though a few since Charlemagne have tried. Yet Europeans from Spain to Sweden have brought from medieval times the idea that however divided they are on earth, still, being Christians, they are all one people on a larger stage.

In medieval times, most people had no time for book-learning. The Church kept it going, but in Latin only. Hardly anyone in Europe knew Greek, and the greatest books of ancient Greece only gradually became known in the twelfth and thirteenth centuries. This was the high point of the Middle Ages, the time of the Gothic cathedrals, of Richard the Lion-Hearted and two great kings of France. It was the time of the Crusades, which didn't work out very well except that the Europeans began to see what a young civilization they had and how much they could learn from the peoples of the Near East, whose history reached back to an age before Latin had been invented.

In the fourteenth century Europe was ruled by the plague—the Black Death—which slowed everyone down. But after that Europeans were ready to take on the Greek heritage, no longer as a monument to be admired, but as a foundation to be continued. This is called the Renaissance or rebirth of Europe, but it was more like a coming of age.

Diophantus

Certain scholars of the Renaissance have remarkably close connections with ancient writers on the same subject. It often seems as though the European had learned from the ancient face to face, not over a gap of fifteen centuries. This is because Greek works dealt with subjects that had not been studied in Europe at all until the Renaissance. Anyone who wanted to learn such a subject had no textbook or teacher nearby. One had to learn it directly from the original Greek masterpiece, and thus come into close contact with the Greek author's mind.

One Greek author who "taught" number theory to the early

Interior of a 16th-century printing shop, from another engraving by Johannes Stradanus, captioned "Just as one voice can be heard by a multitude of ears, so single writings cover a

thousand sheets." After the invention of Gutenberg's printing press in the 15th century, the writings of the ancient world became widely available for Europeans.

European mathematicians was Diophantus of Alexandria, who lived about four hundred years later than Euclid. His book of problems and solutions, called the *Arithmetica*, was studied carefully in the sixteenth century by a mathematician named Raffael Bombelli, and in the seventeenth century by Pierre Fermat. In fact, many of Fermat's discoveries became known through notes that he scribbled on the margin of his copy of the *Arithmetica*. This alone would have made Diophantus famous, but the *Arithmetica* is a wonderful book in its own right, well worth the attention that Bombelli and Fermat gave it.

The first problem in the *Arithmetica* is to find two numbers when you know their sum and difference. Diophantus takes a particular example: the sum is 100, and the difference is 40. He solves it as we would do nowadays. Let x be the smaller number. Then the larger is $x + 40$. The sum is $2x + 40$, which is supposed to equal 100. Therefore $2x = 60$, and $x = 30$. The numbers are 30 and 70.

Obviously, the same method will work if the sum is, for example, 80, and the difference is 30. (Try it out.) But suppose the sum is 100, and the difference is 25. Then we get $2x = 75$, and the numbers are $37\frac{1}{2}$ and $62\frac{1}{2}$. Diophantus has no objection to this; he is perfectly at home with fractions. In fact, he often gives fractional answers to problems for which he could have found solutions without fractions if he had wished. He also does not mind large numbers appearing in fractions. For example, in one problem he arrives at the answer, $4993/784$, $6729/784$, $22660/784$. By this time Greek mathematicians had advanced in the writing of fractions and wrote them very much the way we do.

The freedom to use fractions actually narrowed Diophantus' interests instead of widening them. He paid no attention to problems of divisibility. Nowhere in the *Arithmetica* is there any mention of prime numbers, odd and even numbers, perfect numbers, or anything having to do with the question of whether

“Arithmetic” presiding over a contest between an Abacist and an Algorismist, from the 16th-century book by Gregor Reisch, *Margarita Philosophica.* The Abacist is calculating by means of an old counter or abacus, while the Algorismist uses written numerals. Written numerals based on ancient Hindu-Arabic notation were not in widespread use in schools and trade until the 16th century; before then, abacuses and Roman numerals were most commonly used.

one number goes into another. Diophantus did not need to worry about these matters because he could always divide with the aid of fractions. Thus, in the equation $2x = 75$, it makes no difference to him that 75 is an odd number.

On the other hand, negative numbers were not allowed. If Diophantus arrived at the equation $x + 5 = 3$, he would say, "This is impossible," and try to change the problem so as to get a positive answer. This is usually regarded as a defect in his view of the subject, but its effect was to lead him into problems which he would not have considered if he had been more "advanced." Some of his cleverest tricks were designed to produce a final answer that would not be negative.

This shows that every advance may bring a loss. When people have electric lamps, they may forget how to build a fire. When they have automobiles, they may forget how to ride a horse. When they have weapons they may forget how to make friends, and when they have money they may forget how to pray.

John and James

The next twenty-five problems are all of the kind that we do now in beginning algebra. Here is one, which I have embellished with a story.

John and James make a bet. If John loses, he must give James 30 marbles, and James will then have twice as many as John will. But if John wins, then James must give him 50 marbles, and then John will have three times as many as James will. How many marbles does each boy have to begin with?

Diophantus says let x be the number of marbles that John *will* have left if he loses. Then James will have $2x$ after winning. But if John wins, he will be richer by 80 marbles than if he loses, and James will be poorer by 80. That is, John will have $x + 80$, and James will have $2x - 80$. But John, we are told, will then have three times as many as James, which is $6x - 240$. Therefore $6x - 240 = x + 80$, or $5x = 320$, and $x = 64$. John starts with $x + 30$, or 94 marbles; and James starts with $2x - 30$, or 98.

Notice the clever way of defining x. The obvious thing would be to let x be one of the starting numbers, but then the work would be longer. Also, nowadays we should very likely use two unknowns, x and y—one for John and one for James. Diophantus could not do this because he did not have our system of using letters or made-up symbols, which permits us as many unknowns as we like. He had only one symbol for the unknown, which was an abbreviation of the Greek word for "number." It is as though we used the symbol NR for the unknown.

Having only one symbol for "unknown" did not hamper Diophantus much. He was as nimble with one unknown as Long John Silver on one leg. He always found a way of choosing the unknown quantity so as to do with one what most people would find hard to do with two.

Problem 27 in Diophantus' *Arithmetica* is to find two numbers when you know their sum and product. For example, the sum is to be 20, and the product is to be 96. Let x, says Diophantus, be the amount by which the larger number exceeds 10. That is, let $10 + x$ be the larger number. Then $10 - x$ is the smaller number (so that the sum may be 20) and the product comes out $100 - x^2$, which is supposed to be 96. Therefore $x^2 = 4$, $x = 2$, and the numbers are 12, 8. If we had set one of the numbers equal to x, we should have deduced that $20x - x^2 = 96$, which is not nearly so easy to solve.

If the product is to be 95 instead of 96, and the sum 20 as before, then $x^2 = 5$. This, according to Diophantus, is impossible, even with fractions. He does not allow irrational numbers. That is, he will not write $\sqrt{5}$ and call it a number. The only numbers that exist for him are positive integers and fractions. Consequently, it matters a great deal to him whether or not a number is a perfect square. Most of the problems in the *Arithmetica* have to do with squares. A typical problem, easy for Diophantus, would be to find three squares equally spaced. For example, 1, 25, 49. Or 49, 169, 289. After finishing this chapter, perhaps you will be able to find a formula that gives more solutions.

Choices

We now come to the indeterminate problems. These are problems with more unknowns than conditions, so that one unknown must be given a value at random, and the others deduced. In these problems you don't figure out what the answer must be, the way a detective figures out that no one but Mrs. Ablethwaite could have stolen the jewels. You merely produce a set of numbers which satisfy the conditions, just as an architect produces a plan that satisfies the building code. The plan cannot be reasoned out from the building code. The architect must make choices. But he must make these choices so as not to be led into a violation of the code.

I shall give a few examples that are too easy to have been included in the *Arithmetica*. Find two numbers, given that their sum is 10. All you have to do here is pick any number less than 10, say 8. Then the other number must be 2. Or take 7 for the first number. Then the second number is 3. Notice that the first number is chosen freely, whereas the second is deduced. Another problem is to find two numbers whose product is 10. Here again, you can make the first number anything you like, say 39. Then the second number is $10/39$. Or if you make the first number $2\frac{3}{7}$, then the second is $10 \div 2\frac{3}{7}$ or $4\frac{2}{17}$.

Here is a trickier example. Find two squares which differ by 10. Here you are not free to choose one of the squares. For example, if you choose the lesser square to be 9, then the greater would have to be 19, which is not a square. But a free choice must be made somehow, because there are two unknowns (call them x and y) and only one equation ($x^2 - y^2 = 10$).

The key is to notice that $x^2 - y^2$ is always the product of $x + y$ and $x - y$, no matter what x and y are. You can therefore choose $x + y$ freely, say to be 5. Then $x - y$ has to be $10 \div 5$ or 2. Knowing the sum and difference, you can find x and y as in Diophantus' first problem. The answer is $x = 3\frac{1}{2}$, $y = 1\frac{1}{2}$. The two squares are $12\frac{1}{4}$, $2\frac{1}{4}$, which differ by 10.

Note that a fraction can be a perfect square, if it is the square of another fraction. To find out whether a fraction is a square, first write it as a pure fraction (write $2\frac{1}{4}$ as $\frac{9}{4}$). Then reduce it to lowest terms (reduce $\frac{5}{45}$ to $\frac{1}{9}$). Both the numerator and the denominator must be squares now, if the fraction is a square (thus $\frac{49}{5}$ is not a square). Practice by finding out whether $\frac{9}{60}$, $1\frac{5}{9}$, $1\frac{7}{9}$, $6\frac{3}{12}$ are squares. (Only the last two are.)

The Sum of Two Squares

A harder problem is to find two squares with a given sum. It is harder because $x^2 + y^2$ is not the product of two expressions, as $x^2 - y^2$ is the product of $x + y$ and $x - y$. In fact, it is sometimes impossible, for example if the given sum is 3. Even using fractions, one cannot find x and y such that $x^2 + y^2 = 3$.

The problem can be solved if the given sum is itself a square, and this is the first indeterminate problem in the *Arithmetica*. For example, find two squares that add up to 16. Before I show how Diophantus does it, observe a few methods that don't work!

(a) Choose one square freely, say to be 9. This doesn't work because the remainder is 7, which is not a square.

(b) Choose $x - y$ freely, say to be 2. Then $y = x - 2$, and

$x^2 + y^2 = x^2 + x^2 - 4x + 4 = 2x^2 - 4x + 4$. So x must be deduced from the equation $2x^2 - 4x + 4 = 16$. Besides being hard to solve, this equation leads not to a fractional answer, but to irrationals, which are not allowed.

(c) Choose $x + y$ freely, say to be 5. Then $y = 5 - x$, and $2x^2 - 10x + 25 = 16$. This also leads to an irrational number.

Now Diophantus' solution: x^2 must be less than 16, so that x is less than 4. So x is certainly less than $y + 4$. Choose freely the *ratio* of x to $y + 4$, only making sure that it is less than 1. For example, choose x to be half of $y + 4$. Then $y = 2x - 4$, and $y^2 = 4x^2 - 16x + 16$. Subtracting this from 16, we find that $x^2 = 16x - 4x^2$. Therefore $5x^2 = 16x$, or $5x = 16$, and $x = 16/5$, $y = 12/5$. The two squares are $10\,6/25$, $5\,19/25$.

It seems like magic. We could also choose $y = 3x - 4$. Then $y^2 = 9x^2 - 24x + 16$, $x^2 = 24x - 9x^2$, and $x = 24/10 = 12/5$, $y = 16/5$, which is the same solution reversed. Or $y = 5x - 4$, $y^2 = 25x^2 - 40x + 16$, $x^2 = 40x - 25x^2$, $x = 20/13$, $y = 48/13$. The two squares are then $2\,62/169$, $13\,107/169$.

Any number greater than 1 can be used to multiply x. But the number subtracted must be 4. Thus $y = 2x - 3$ will not work, for then $y^2 = 4x^2 - 12x + 9$, $x^2 = 12x - 4x^2 + 7$, and x is irrational. The whole solution depends on the cancellation of 16 so that the final equation involves only x and x^2. Then it can be solved without square roots.

Diophantus' next problem is to deal with a number which is not a square but is the sum of two squares, such as 13, which is $4 + 9$. He wants to find two *other* squares which add up to 13. His method is very much like the one for 16. He arranges to cancel the 13. The simplest way would be to let $x = 2 + u$, and $y = 3 - u$. Then

$$\begin{aligned} x^2 + y^2 &= 4 + 4u + u^2 + 9 - 6u + u^2 \\ &= 13 - 2u + 2u^2 \end{aligned}$$

This is supposed to equal 13. Therefore $2u^2 = 2u$, and $u = 1$. So far, so good, but now $x = 3$ and $y = 2$, which gives the original squares in reverse.

Therefore Diophantus lets $x = 2 + u$ and $y = 2u - 3$, explaining that he could just as well use $3u - 3$, $4u - 3$, etc. He gets

$$\begin{aligned} x^2 + y^2 &= 4 + 4u + u^2 + 9 - 12u + 4u^2 \\ &= 13 - 8u + 5u^2 \end{aligned}$$

so that $5u^2 = 8u$, $u = 8/5$, $x = 18/5$, and $y = 1/5$. The two squares are $12\,24/25$ and $1/25$. See if you can use this method to find two squares that add up to 10, besides 9 and 1.

The Neighborhood Square

Diophantus uses cancellation in a different way in order to find a square very close to a given number. He wants a square very near $6\frac{1}{2}$. He knows that $6\frac{1}{4}$ is a square, but this is not close enough. So, he says, consider the square of $x + 2\frac{1}{2}$. It is $6\frac{1}{4} + 5x + x^2$. The idea now is to put some condition on x that makes x rational and causes $5x + x^2$ to be close to $1/4$. The easiest way to do this is by letting $5x = 1/4$, or $x = 1/20$. Then $x + 2\frac{1}{2} = 51/20$, and the square of $51/20$ is $6\frac{1}{2} + 1/400$.

This procedure can be shortened into a rule of thumb. Given that the square of $2\frac{1}{2}$ is close to $6\frac{1}{2}$, then $6\frac{1}{2} \div 2\frac{1}{2} = 13/5$ must be close to $2\frac{1}{2}$. Take the average of $2\frac{1}{2}$ and $13/5$, which is $51/20$. Its square is even closer to $6\frac{1}{2}$.

The process can be repeated, starting with $51/20$; $6\frac{1}{2} \div 51/20 = 130/51$. The average of $51/20$ and $130/51$ is $5201/2040$, and its square is $6\frac{1}{2} + 1/4{,}161{,}600$. This is a fine method for finding square roots to many decimal places, since it works better, the closer one is to the true root. The method taught in school becomes slower and slower, as one goes to more decimal places.

As an example, take the square root of 2 by the usual school method.

```
      |  2. 00 00 00 00    |1.4142
      |  1
      |  -
      |  1 00
   24 |    96
      |  ----
      |     4 00
  281 |     2 81
      |     ----
      |     1 19 00
 2824 |     1 12 96
      |     -------
      |        6 04 00
28282 |        5 65 64
      |        -------
      |          38 36
```

So far the work is not difficult, but the next five decimal places would be harder. Let us go on with Diophantus' method instead. The remainder on the bottom line means that $1.4142^2 = 2 - .00003836$. Therefore $\frac{2}{1.4142} = 1.4142 + \frac{.00003836}{1.4142}$. The average of this with 1.4142 is $1.4142 + \frac{.00001918}{1.4142}$. Its square is $2 + \left(\frac{.00001918}{1.4142}\right)^2$ or about $2 + .0000000002$. Therefore the square root of 2 is $1.4142 + \frac{.00001918}{1.4142}$ minus a small error in the tenth decimal place.

To avoid dividing by 1.4142, we may write $\frac{.00001918}{1.4142} = \frac{1.4142 \times .00001918}{2 - .00003836}$ and drop the little term in the denominator. This makes an error in the tenth decimal place, which was in error anyway. Therefore the square root of 2 is 1.4142 × 1.00000959, correct to nine places. Note that 1.00000959 is found by dividing .00003836 by twice 2 and adding 1. The work can be shown thus:

```
          |  2. 00 00 00 00  |1.4142
          |  1
          |  1  00
      24  |     96
          |      4 00
     281  |      2 81
          |      1 19 00
    2824  |      1 12 96
          |         6 04 00
   28282  |         5 65 64
twice 2 →        4 )38 36              short division
             1 .00 00 09 59
            ×1 .41 42                  long multiplication
             1 .00 00 09 59
               .40 00 03 836
               .01 00 00 0959
               .00 40 00 03836
               .00 02 00 001918
             1 .41 42 13 562178        keep nine decimals
```

In this way the number of decimal places is doubled by one long multiplication. Try checking this answer by carrying the school method to nine places. Also try the method on 3. Here you must divide the remainder by 6, instead of 4.

Over a Gate

Diophantus surpasses himself when he sets out to find two squares, *nearly equal* to each other, which add up to 13. He knows that if m is any number at all, then without taking square roots he can find another number u, so that the squares of $2 + u$ and of $3 - mu$ add up to 13. This will be done by cancelling 13 in an equation, as I have described already. The problem now is to choose m in the first place, so that after u is found it will turn out that $2 + u$ and $3 - mu$ are nearly equal. (Diophantus manages to explain all this by using only one symbol for an unknown quantity, but I can make it clearer by using two.)

To choose m, Diophantus first finds a square slightly larger than $6\frac{1}{2}$, by the method I have already explained. He finds $(51/20)^2 = 6\frac{1}{2} + 1/400$. Therefore, he reasons, if $2 + u$ and $3 - mu$ were both equal to $51/20$, he would have two squares *exactly* equal and adding up to *nearly* 13. Then by changing u a little bit, he would find two squares *nearly* equal and adding up to *exactly* 13.

Thus he first chooses u to make $2 + u = 51/20$. This gives $u = 11/20$. Knowing u, he then chooses m to make $3 - mu = 51/20$, or $3 - 11/20m = 51/20$. This gives $m = 9/11$. Now he keeps his m but throws away his u, and looks for a new u satisfying

$$(2 + u)^2 + (3 - 9/11u)^2 = 13$$

or

$$4 + 4u + u^2 + 9 - 54/11u + 81/121u^2 = 13$$

or

$$202/121u^2 = 10/11u$$

or

$$u = 55/101,\ 2 + u = 257/101,\ 3 - 9/11u = 258/101$$

The problem is now solved, since $(257/101)^2 + (258/101)^2 = 13$. This method reminds me of the White Knight's invention for getting over a gate in *Through the Looking Glass.* First you lay your head on the gate; then the head is high enough. Next you stand on your head, that makes the feet high enough—and then . . .

5.

A narrow margin

Right Triangles

Diophantus devotes a whole chapter to right-angled triangles. This subject had belonged to number theory ever since Pythagoras found out that the square of the length of the hypotenuse (the long side) is always equal to the sum of the squares of the two legs (the two shorter sides). This is a geometrical theorem, true even if the lengths are not rational fractions of one another. But it shows that in certain cases the lengths *are* commensurate, that is, all three sides can be marked off as exact multiples of some one unit. For example, if one leg is 3 units long, and the other 4 units, then the hypotenuse is exactly 5 units long, since $3^2 + 4^2 = 5^2$.

If you select the legs at random, the hypotenuse will usually be incommensurate with them. For example, if the legs are equal, then the square of the hypotenuse is twice that of a leg. But the equation $2x^2 = y^2$ cannot be satisfied by integers, as proved in Chapter 2. This does not mean, of course, that a right triangle cannot be drawn with two equal legs, only that its sides cannot all be measured as whole numbers or fractions.

Egyptian surveyors had long known about the triple (3, 4, 5), and had used it to mark off a right angle with three ropes of the right lengths. But after Pythagoras' discovery, the Greeks found other combinations that would do the job. For example,

$$5^2 + 12^2 = 13^2$$
$$15^2 + 8^2 = 17^2$$
$$7^2 + 24^2 = 25^2$$
$$21^2 + 20^2 = 29^2$$

Euclid's *Elements* gives a rule for making as many "Pythagorean triplets" as you like. The idea is that since

$$(u + v)^2 = u^2 + 2uv + v^2$$

and

$$(u - v)^2 = u^2 - 2uv + v^2$$

it follows that

$$(u + v)^2 = (u - v)^2 + 4uv$$

Now if u and v are both squares to begin with, then $4uv$ is a square, and we have a Pythagorean triplet. In other words,

$$(m^2 + n^2) = (m^2 - n^2)^2 + (2mn)^2$$

no matter what m and n are. So choose any pair of integers, call them m and n, and let

$$x = m^2 - n^2$$
$$y = 2mn$$
$$z = m^2 + n^2$$

Then (x, y, z) is a Pythagorean triplet. ("Pythagorean" rhymes with "Why drag a bee in.")

If $m = n$, then $x = 0$, which doesn't count. So the lowest values for (m, n) are $(2, 1)$. Then

$$x = 2^2 - 1^2 = 3$$
$$y = 2 \times 2 \times 1 = 4$$
$$z = 2^2 + 1^2 = 5$$

Try a few other pairs for (m, n), such as (3, 1), (3, 2), (4, 1), (5, 2), (4, 3), (5, 3), (6, 3). You will see that sometimes a triangle is repeated in higher multiples. For example, from (3, 1) we get

$$x = 3^2 - 1^2 = 8$$
$$y = 2 \times 3 \times 1 = 6$$
$$z = 3^2 + 1^2 = 10$$

But (6, 8, 10) gives the same shape triangle as (3, 4, 5). On the other hand, from (3, 2) we get (5, 12, 13), which is a new shape.

The triangle (9, 12, 15) cannot be obtained from any pair (m, n), because 15 is not the sum of two squares. On the other hand, this triangle is just (3, 4, 5) repeated in multiples of 3. A good question is: can all *primitive* right triangles be made from pairs (m, n)? (A right triangle is primitive if it is not the same shape as any smaller triangle.)

This question was not really explored by the Greeks, who were better at synthesis than at analysis. That is, they were better at putting things together than taking them apart. They discovered how to put m and n together to make a right triangle. But we wish now to take apart a primitive right triangle and reveal its key numbers m and n. The technique for this sort of problem is based on the unique factorization theorem, and it was used a great deal by Euler and was certainly known to Fermat. But it is not found in Euclid or Diophantus, although the unique factorization theorem *itself* is essentially demonstrated in the *Elements*.

Taking Apart a Square

Let me introduce the technique by proving two theorems about squares.

First theorem. If a square number is factored into primes, $n^2 = p^a q^b r^c \ldots$, then all the exponents $a, b, c, \ldots$ are even.

This is obvious, because n can be factored somehow, say $n = p^u q^v r^w \ldots$, and then we have $n^2 = p^{2u} q^{2v} r^{2w} \ldots$. There is no other way of factoring n^2, because of the unique factorization theorem.

If the early Pythagoreans had understood this theorem, they would have seen much more easily that the double of a square cannot be a square. If the smaller square has the factor 2^a, then its double has 2^{a+1}. But a and $a + 1$ cannot both be even.

Second theorem. If m and n have no factor in common, and mn is a square, then m and n are both squares.

This is obvious if the first theorem is granted. For let $m = p^a q^b r^c \ldots$, and $n = s^x t^y u^z \ldots$. Then the primes $p, q, r \ldots$ are all different from $s, t, u \ldots$ since m and n have no factor in common. Therefore $mn = p^a q^b r^c \ldots s^x t^y u^z \ldots$, and all the exponents $a, b, c, \ldots, x, y, z, \ldots$ are even, because of the first theorem. Hence m and n are both squares. (Do you see why?)

Take $6^2 = 36$, for example. It can be factored into two numbers in several ways, but the two numbers always have some common factor except when $36 = 36 \times 1$ or $36 = 4 \times 9$. In both these cases the two numbers are squares.

Finding the Key

Now we may analyze a primitive right triangle, $x^2 + y^2 = z^2$. First note that x and y cannot both be odd, because the sum of two odd squares always contains the factor 2 to the first power and no more. To prove this, let $x = 2k + 1$, $y = 2j + 1$. Then $x^2 + y^2 = 4k^2 + 4k + 4j^2 + 4j + 2 = 2$ times an odd number. This cannot be a square.

On the other hand, x and y cannot both be even, for then z would be even and the triangle would not be primitive. It would be similar to another triangle with sides half as long. Hence one leg is odd and the other even, and we may decide to call the odd leg x, and the even leg y.

Now since z and x are both odd, their sum and difference are both even. So we can divide them both by 2, and write

$$u = \tfrac{1}{2}(z + x), \qquad v = \tfrac{1}{2}(z - x)$$

which implies that $z = u + v$, $x = u - v$. Then $y^2 = (u + v)^2 - (u - v)^2 = 4uv$, so that uv is a square.

If u and v had a common prime factor, their sum and difference, z and x, would also have this factor. But then y would have it also, since $y^2 = z^2 - x^2$. This is impossible since the triangle is primitive. Therefore we can use the second theorem above and deduce that u and v are both squares. Thus

$$u = m^2, \qquad v = n^2$$

and $$x = m^2 - n^2, \qquad y = 2mn, \qquad z = m^2 + n^2$$

Therefore every primitive right triangle has a pair of key numbers (m, n).

Key numbers are a great help in finding particular kinds of Pythagorean triplets. In the triplets (3, 4, 5), (5, 12, 13), (7, 24, 25), the hypotenuse is just 1 more than one of the legs. How can we find other triplets of this kind?

Since $z - y = m^2 + n^2 - 2mn = (m - n)^2$, all we need do is choose (m, n) to be two consecutive numbers. For example, from (4, 3) we get (7, 24, 25), from (5, 4) we get (9, 40, 41), from (6, 5) we get (11, 60, 61).

A Slight Limp

A harder problem is to make the two legs differ by 1. Two solutions are (3, 4, 5) and (21, 20, 29).

Using key numbers, we see that

$$\begin{aligned} x - y &= m^2 - n^2 - 2mn \\ &= m^2 - 2mn + n^2 - 2n^2 \\ &= (m - n)^2 - 2n^2 \end{aligned}$$

so that the problem is to find two *consecutive* numbers, of which one, $(m - n)^2$, is a square and the other, $2n^2$, is double a square. The smallest such pair is $(m - n)^2 = 1$, $2n^2 = 2$. This yields $m = 2$, $n = 1$, and $(x, y, z) = (3, 4, 5)$. The next smallest is $2n^2 = 8$, $(m - n)^2 = 9$. Then $m = 5$, $n = 2$, $(x, y, z) = (21, 20, 29)$.

To find more solutions, we can treat $x - y$ a little differently.

$$\begin{aligned} x - y &= m^2 - n^2 - 2mn \\ &= 2m^2 - m^2 - 2mn - n^2 \\ &= 2m^2 - (m + n)^2 \end{aligned}$$

This means that from the original triangle we can find a *second* pair of consecutive numbers, of which one is a square and the other double a square. The second pair is larger than the first, and it can be used to find a second right triangle, larger than the first, whose legs also differ by 1.

For example, starting from (21, 20, 29), with $m = 5$, $n = 2$, we have $2m^2 = 50$, $(m + n)^2 = 49$. To get a larger solution, let $2j^2 = 50$, $(k - j)^2 = 49$. That is, $j = 5$, $k - j = 7$, $k = 12$. Then (12, 5) are the key numbers for a new triangle, (119, 120, 169). The legs of this triangle differ by 1. We now have $2k^2 = 288$, $(k + j)^2 = 289$, and these can be used to make a still larger triangle.

This procedure can be continued indefinitely. If we don't care about the triangles, but only want to find pairs (a, b) such that a^2 and $2b^2$ are consecutive, then the method is quite simple. We need only start with one solution, say (a_0, b_0), and find (m, n) so that $a_0 = m - n$, $b_0 = n$. Then a new solution (a_1, b_1) is given by $a_1 = m + n$, $b_1 = m$. In fact, m and n can be eliminated since obviously $b_1 = a_0 + b_0$, $a_1 = b_1 + b_0 = a_0 + 2b_0$.

It is easy to see that this works. We have

$$a_1{}^2 = a_0{}^2 + 4a_0b_0 + 4b_0{}^2$$
$$2b_1{}^2 = 2a_0{}^2 + 4a_0b_0 + 2b_0{}^2$$

and therefore $a_1{}^2 - 2b_1{}^2 = 2b_0{}^2 - a_0{}^2$.

In the same way we can put $b_2 = a_1 + b_1$, $a_2 = a_1 + 2b_1$, and so on. If we start with $a_0 = 1$, $b_0 = 0$, then the successive pairs are (1, 1), (3, 2), (7, 5), (17, 12), (41, 29), and so on. Each pair represents an isosceles triangle which is nearly right-angled: thus $29^2 + 29^2 = 841 + 841 = 1682$, while $41^2 = 1681$. From two consecutive pairs we may take the smaller partners as key numbers to make a right triangle which is nearly isosceles: thus from (29, 12) we make (697, 696, 985).

In Euclid's *Elements* this sequence of pairs is mentioned, along with the rule for generating it

$$a' = a + 2b$$
$$b' = a + b$$

The rule is supposed to have been known since the time of Pythagoras. Since the Greek mathematicians then were much handier at geometry than at algebra, it is likely that they discovered it by some geometrical method. Here is one, which they were quite clever enough to have used.

A Little Geometry

Take an isosceles right-angled triangle (OAB). Draw another triangle ($OA'B'$), the same shape and size as the first, but flipped about the angle at O. The sides AB and $A'B'$ intersect at C.

Because of the symmetry of the figure, we know that $\overline{OA} = \overline{OA'}$, $\overline{AB} = \overline{A'B'}$, $\overline{BC} = \overline{B'C}$, etc.

Suppose that $\overline{OA}$ and $\overline{OB}$ could be marked off as exact multiples of some unit. Then $\overline{OB'}$ could be marked off (equal to $\overline{OB}$) and therefore the remainder $\overline{AB'}$ could be marked off. But the angle at A is 45°. Therefore $\overline{B'C} = \overline{AB'}$, and of course $\overline{A'B'} = \overline{OB'}$. So $\overline{A'B'}$ and $\overline{B'C}$ could be marked off, and also $\overline{A'C}$, their difference. Therefore every length on the diagram could be marked off exactly in multiples of one unit. Since $(AB'C)$ forms an isosceles right triangle smaller than (OAB), we could draw a new figure starting with $(AB'C)$. Thus we could arrive at smaller and smaller isosceles right triangles, each expressible in smaller multiples of the same unit. This is impossible, being an infinite descent, and we now have a third proof that the sides of an isosceles right triangle are incommensurate.

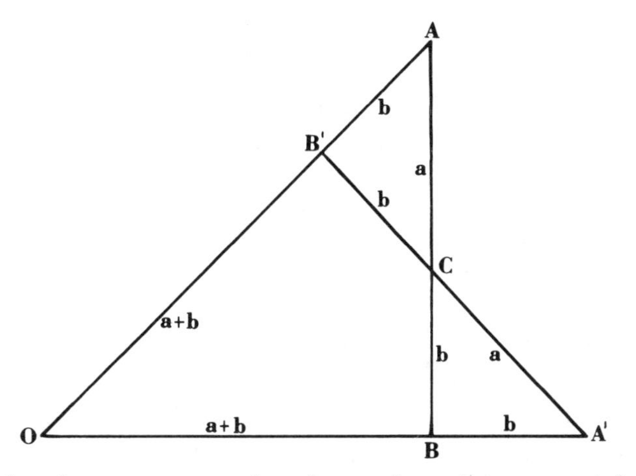

Now let us suppose that the angle at B is *not* a right angle. If we write a for the length of $\overline{AC}$, and b for $\overline{BC}$, then the diagram shows that $a + b$ is the length of $\overline{AB}$, and $b + (a + b)$ or $a + 2b$ is the length of $\overline{OA}$. Therefore, if (a, b) are the longer and shorter sides of an isosceles triangle which is *nearly* right-angled, then $(a + 2b, a + b)$ form a larger isosceles triangle which is nearly the same shape. But which triangle is closer to a right triangle?

Lay part of the figure out as before, so that $\overline{AC} = \overline{A'C} = a$, $\overline{AB'} = \overline{B'C} = \overline{BC} = \overline{A'B} = b$, only let the angle $AB'C$ be more than a right angle. Then $\overline{OB}$ is shorter than $\overline{AB}$, as can be seen at a glance and proved by studying the angles.

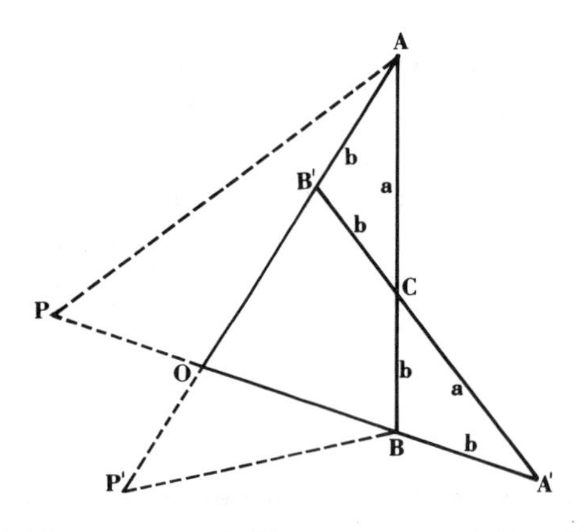

Extend $\overline{AO}$ to P' and $\overline{A'O}$ to P so that $\overline{BP} = \overline{B'P'} = a + b$. Then $\overline{BO} + \overline{OP'} = \overline{B'O} + \overline{OP'} = a + b$. Hence $\overline{BP'}$ is less than $a + b$. By the same reasoning, $\overline{AP}$ is less than $a + 2b$.

Now find a point Q such that (ABQ) is an isosceles triangle with sides $(a + 2b, a + b)$. That is, $\overline{BQ} = a + b$, which is equal to $\overline{BP}$ and greater than $\overline{BP'}$; and $\overline{AQ} = a + 2b$, which is equal to $\overline{AP'}$ and greater than $\overline{AP}$.

From these inequalities it follows that Q lies somewhere between P and P'. In particular, the angles QBA and QAB are greater than the angles OBA and OAB, respectively. From this it can be proved that the angle QBA is closer to a right angle than the angle $AB'C$. Therefore the shape of the triangle improves as we make it larger, going from (a, b) to $(a + 2b, a + b)$ and so on.

It is true that this reasoning accomplishes less, with more

effort, than the algebraic method. But it is more likely to have been discovered by the early Greek mathematicians.

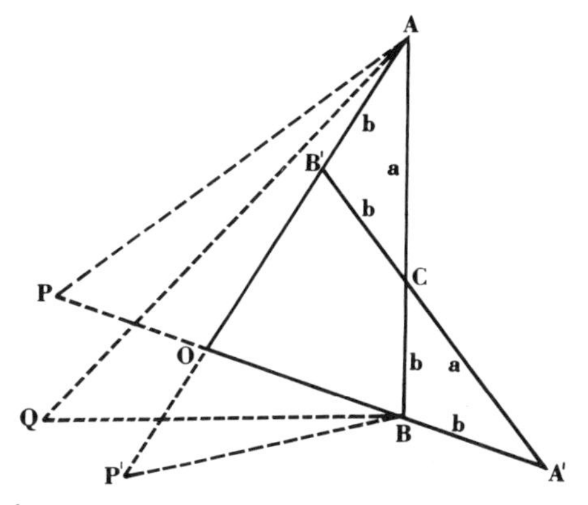

Along the Chain

The chain of triangles can be run downward instead of upward. Given (a, b) we find a smaller pair $(2b - a, a - b)$. Thus, from (17, 12) we find (7, 5).

Why doesn't this lead to an infinite descending sequence? Well, when it reaches bottom, it stops descending. From (1, 1), we find (1, 0), and from (1, 0), we find $(-1, 1)$. The sequence then *ascends*, with negative numbers. Or, if you don't allow negative numbers, the sequence stops at (1, 0) since $2b$ is less than a.

This reasoning shows that the sequence we have found contains *all* pairs (a, b), such that a^2 and $2b^2$ are consecutive. For if any such pair has b greater than 1, it is easy to see that a lies between $2b$ and b. Therefore a chain can be run downward to $(2b - a, a - b)$, which are both positive and less than (a, b). As long as the new b exceeds 1, another downward step is possible. Since downward steps cannot continue forever, one must arrive eventually at $b = 1$ or 0. But if $b = 1$ or 0, a must be 1. Therefore the chain in question is the very one we have studied. There are no other chains.

No Such Animal

The Greek authors did not often use the method of infinite descent, although they knew of it, as shown by the proof I described at the end of Chapter 2. Being synthesists rather than analysts, they did not attach much value to proving that a thing was impossible. Fermat took a different attitude and laid great stress on proofs of impossibility. His point of view gradually infected all of mathematics, so that nowadays it is considered quite as much of an achievement to prove that a problem cannot be solved as to solve it! But to the Greeks, such a proof was a disaster rather than an achievement. It was as though a painter were to prove that a picture could not be painted, or a general, that a battle could not be won. Even when they knew a problem was impossible, the Greeks usually displayed their knowledge by avoiding the problem, rather than by announcing that it could not be solved.

Diophantus showed how to find many special right triangles. For example, the hypotenuse and the difference of two legs should be squares. (Answer: 119, 120, 169.) Or the area plus the hypotenuse should be a square, and the perimeter should be a cube. (Answer: 2, $621/50$, $629/50$.) But Fermat noticed that Diophantus had not mentioned what seemed to be an easier problem: finding a right triangle with its area equal to a square. Fermat therefore set out to find whether this could be done.

The area of a right triangle is half the product of the two legs. Using key numbers, we find

$$\begin{aligned} A = \tfrac{1}{2}xy &= \tfrac{1}{2}(m^2 - n^2)(2mn) \\ &= mn(m^2 - n^2) \end{aligned}$$

Therefore the problem is to find m, n such that $mn(m^2 - n^2)$ is a square. One may as well require m, n to be integers, for if they are fractions, they can be multiplied by their common denominator, and the area will still be a square.

Fermat's first step was to apply the same kind of reasoning that we used in showing that a primitive right triangle always has key numbers. If m and n had a common factor, then x, y, and z would have it too. This factor could be removed, and the area of the new triangle would still be a square. Therefore we may assume that m and n have no common factors. Nor do m and $m^2 - n^2$ have a common factor, for then n^2 would have it, and n would have some factor of it. Thus no two of the three numbers $m, n, m^2 - n^2$ have a factor in common. Hence $m, n, m^2 - n^2$ are all squares.

Now m and n can be chosen as square, and then $m^2 - n^2$ is the difference between two fourth powers. Therefore the original problem can be solved only by finding two fourth powers whose difference is a square. Fermat looked this problem up in Diophantus, and found that it was not mentioned, although there were several problems on fourth powers that seemed more difficult on the surface. This deepened his suspicion that the original problem could not be done.

The quantity $m^2 - n^2$ is equal to $(m + n)(m - n)$. Now $m^2 - n^2$ must be odd, if the original triangle is primitive. Therefore both $m + n$ and $m - n$ are odd. This means that if they had any factor in common, m and n would have it too (for $2m$ and $2n$ are the sum and difference of $m + n$ and $m - n$, and the factor would have to be odd). Hence $m + n$ and $m - n$ have *no* common factor and they both must be squares. Now all four numbers, $m, n, m + n, m - n$, are known to be squares.

There is a clever trick often used by Diophantus when the sum and difference of two numbers are both to be made squares. It is to let $m + n$ be the square of $a + b$, and $m - n$ the square of $a - b$. This is always possible, because if $m + n = u^2$ and

$m - n = v^2$, we can let $a = \frac{1}{2}(u + v)$ and $b = \frac{1}{2}(u - v)$. In this problem, $m + n$ and $m - n$ are both odd. Therefore $u + v$ and $u - v$ are both even, and a and b are integers.

Letting

$$m + n = (a + b)^2 = a^2 + 2ab + b^2$$
$$m - n = (a - b)^2 = a^2 - 2ab + b^2$$

we see that

$$m = a^2 + b^2$$
$$n = 2ab$$

But m and n are also squares

$$c^2 = a^2 + b^2$$
$$d^2 = 2ab$$

Therefore (a, b, c) is a right triangle, and its area is $\frac{1}{2}ab = \frac{1}{4}d^2$, which is the square of $\frac{1}{2}d$. Thus (a, b, c) is a second solution of the original problem.

When Fermat arrived at the triangle (a, b, c), he realized that the problem was impossible, for c is less than the original hypotenuse z. If the whole argument were repeated with (a, b, c) as the starting point, it would lead to a third triangle whose area is also a square, and so on forever. The hypotenuses would make an infinite descending sequence.

We must make sure that c is always less than z. Actually, $z = m^2 + n^2 = c^4 + n^2$. Therefore c is less than z, unless $c = 1$ and $n = 0$. In that case, (a, b, c) is $(1, 0, 1)$. This triangle does satisfy the theorem since its area is 0. But a zero solution is not allowed. No higher solution exists because there is no chain running upward from $(1, 0, 1)$. If $c = 1$ and $n = 0$, we find that (x, y, z) is $(1, 0, 1)$, the same as (a, b, c). And there cannot be a chain that does not include $(1, 0, 1)$, because it would lead to an infinite descent. This problem differs from the one of making a^2 and $2b^2$ consecutive numbers. There the pair (1, 0) led upward to (1, 1), (3, 2), and a chain of higher solutions. Here the triangle (1, 0, 1) leads only to itself, so that there are no higher solutions.

Fermat explained this proof in a letter—a thing he rarely did. He wanted to make clear the idea of infinite descent, which seems to have been the main tool by which he proved difficult theorems. He referred to it often in announcing theorems "which I have shown by my method of infinite descent." Unfortunately he didn't usually explain how the descent was achieved.

Since there cannot be an infinite descent, no link in the chain can exist. Not only is there no right triangle with area equal to a square, but there are no two squares (m and n) whose sum and difference are both squares, and there are no two fourth powers (m^2 and n^2) whose difference is a square, for any one of these things would start a downward chain.

Any fourth power is a square. Therefore a fourth power cannot be the sum of two fourth powers. If $x^4 + y^4 = z^4$, then we can let $m = z^2$, $n = x^2$, and $m^2 - n^2 = y^4$. Then m, n, and $m^2 - n^2$ are all squares, which we have proved to be impossible.

A Narrow Margin

This discovery must have set Fermat to wonder whether a cube can be the sum of two cubes. He seems to have solved this problem, and made a note of it on the margin of his copy of Diophantus' *Arithmetica.* (After his death, this copy was found and a special edition was printed that contained all of his marginal notes.) Next to the problem about finding two squares that add up to 16, Fermat wrote, "It is impossible to partition a cube into two cubes or a fourth power into two fourth powers, or generally any power higher than the second into two powers

Pierre de Fermat, 1601–1665

of the same degree. I have discovered a truly wonderful proof of this, which, however, this margin is too narrow to contain."

To prove that a cube cannot be the sum of two cubes is harder than to prove the same thing for fourth powers. It was finally done (see Chapter 7) by Euler, using a method of infinite descent that resembles the one I have described here, except that it requires some additional knowledge about the expression $p^2 + 3q^2$. Now, Fermat did know quite a lot about this expression, as shown by some of his letters. So he may have proved the statement about cubes by the same method as Euler.

But how could he have proved the whole theorem? The proof for fifth powers is more complicated than that for cubes, although Fermat might conceivably have discovered it. But similar methods just don't work for the seventh or higher powers. In the nineteenth century a proof was discovered that $x^n + y^n = z^n$ is impossible for a very large class of values of n greater than 2. In the twentieth century even more values of n were excluded, and it is now known that the smallest value of n for which Fermat's statement could possibly be wrong is 619. To this day, the complete theorem has not been proved, even though the methods now in use are much more advanced than anything Fermat knew.*

If Fermat was wrong, there lie in wait two enormous numbers, both powers of the same degree, which add up to a power of the same degree; and if we should ever chance on them the problem would be solved. This seems unlikely from what has already been proved—but remember the examples in Chapter 2 and don't jump to conclusions!

The Purloined Letter

Fermat's reputation was very high among mathematicians of his own time. It grew higher in the next century, when Euler and

* See note in Appendix 1.

others proved again and again that Fermat's statements were correct. By the time he had been dead for two hundred years, a proof had been published for every one of the theorems Fermat claimed to have proved, except this one about the sum of two nth powers. It became known as "Fermat's Last Theorem," since it was the only one remaining unsolved.

Most mathematicians have thought all along that Fermat's Last Theorem is true—after all, no one had proved it false—but few believe it possible that Fermat had proved it. He was certainly ahead of his own time, but he was not ahead of ours. And yet he claimed to have proved it. Could he have been mistaken?

You may remember that Fermat *was* mistaken about another point, when he thought that the numbers 3, 5, 17, 257, 65537, etc., would all be prime. But he was only guessing about that. He did not say he had proved it. If he was mistaken about the Last Theorem, it was more than a mistaken guess; he must have discovered an argument which he thought was air-tight, but which really had a flaw. This is not impossible. Many mathematicians as great as Fermat have published "proofs" of theorems which contained subtle errors. But then it is no easier to see what his false proof could have been than it is to find a true proof. No one has discovered any line of reasoning that Fermat might have used, which even could be mistaken for a proof of his Last Theorem.

The situation reminds me of Edgar Allan Poe's story, *The Purloined Letter.* The hero, an early model of Sherlock Holmes, is consulted by a police inspector who is trying to recover an important letter. He knows the letter was stolen by a certain cabinet minister, but his agents cannot find it when they make secret searches. He describes in detail the fantastic precautions he has taken to discover any hidden compartment in the minister's home—thrusting needles up the legs of chairs, measuring all the rooms from inside and outside the house, and so on. Dupin, the hero, visits the evil minister's home and spots the letter in plain sight, where the minister had been sure that the police would overlook it.

Afterward, Dupin remarks that he felt certain from the start that it would do no good to intensify the police inspector's methods by probing the chair legs twice and measuring the walls more precisely. If the inspector had not already found the letter, then it could not be found *by his methods.* In the same way it seems that whether Fermat's proof was correct or faulty, it will not be discovered by the modern methods, which are much too advanced for Fermat to have used. He must have taken another direction.

There is evidence that Fermat did take directions that were different from anything discovered later. One of his theorems, which I haven't room to describe here, was a puzzle until the nineteenth century, when it was proved by the French mathematician Augustin Louis Cauchy. But Cauchy's proof is certainly not the one that Fermat had found. It required knowledge that Fermat didn't have, as shown by his letters. It also leads to a stronger theorem, which Fermat would have mentioned if he had known it. And in one letter, Fermat gave some hints as to his method of proof, which show that it was not at all like Cauchy's. This method of Fermat's is still undiscovered.

The Primitive Hypotenuse

One of Fermat's finest achievements was to find out which numbers are the hypotenuse of a primitive right triangle. Using key numbers, you can easily see that a "primitive hypotenuse" is always odd and equal to the sum of two squares with no common factor. (The *square* of the hypotenuse is the sum of two squares, thus $5^2 = 3^2 + 4^2$, but the hypotenuse *itself* is also the sum of two squares, thus $5 = 2^2 + 1^2$. Keep this straight.) By trying out key numbers, you can start a list of primitive hypotenuses: 5, 13, 17, 25, 29, 37, 41, 53, 61, 65, 73, etc. The problem is to find a rule (an inclusive rule, in the language of Chapter 2) which describes this list.

First off, you will notice that the intervals between numbers on the list are 4, 8, 12, but never 2, 6, etc. In other words, every primitive hypotenuse is one more than a multiple of 4. Let me call these numbers, 1, 5, 9, 13, etc., "upper numbers," and let the "lower numbers," 3, 7, 11, etc., be those which are 1 less than a multiple of 4.

It is easy to prove that every primitive hypotenuse is an upper number. The sum of an even square and an odd square can be written as

$$\begin{aligned}(2a)^2 + (2b+1)^2 &= 4a^2 + 4b^2 + 4b + 1\\ &= 4(a^2 + b^2 + b) + 1\end{aligned}$$

But which upper numbers are primitive hypotenuses?

Most of the numbers on the list are prime, but not all. In fact, Diophantus knew that the product of two primitive hypotenuses is a hypotenuse. This can be shown by some dazzling algebra:

$$\begin{aligned}(a^2 + b^2)(c^2 + d^2) &= a^2c^2 + a^2d^2 + b^2c^2 + b^2d^2\\ &= (a^2c^2 + 2abcd + b^2d^2) +\\ &\quad\ (a^2d^2 - 2abcd + b^2c^2)\\ &= (ac + bd)^2 + (ad - bc)^2\end{aligned}$$

By interchanging c and d, we get a second way of making the product a hypotenuse:

$$(a^2 + b^2)(c^2 + d^2) = (ac - bd)^2 + (ad + bc)^2$$

For example, from

$$5 = 2^2 + 1^2$$

and

$$13 = 3^2 + 2^2$$

we get

$$\begin{aligned} 5 \times 13 &= (2 \times 3 + 1 \times 2)^2 + (2 \times 2 - 1 \times 3)^2 \\ &= (2 \times 3 - 1 \times 2)^2 + (2 \times 2 + 1 \times 3)^2 \end{aligned}$$

or

$$\begin{aligned} 65 &= 8^2 + 1^2 \\ &= 4^2 + 7^2 \end{aligned}$$

I shall not bother to prove that there is always some way of choosing the two final squares with no common factor, so that the product is actually a primitive hypotenuse.

There are now two problems. One is to find out which *primes* are primitive hypotenuses. Then any number which has only such primes as factors will be a primitive hypotenuse. Only upper primes can qualify, but which ones?

The second problem is to find out which numbers are primitive hypotenuses, even though some of their prime factors are not. This list will include only upper numbers, but they may have lower prime factors, since the product of two lower numbers is an upper number.

Fermat announced the solution of both problems: *every* upper prime is a primitive hypotenuse, and *no* primitive hypotenuse has a lower prime factor. Therefore a number is a primitive hypotenuse if, and only if, *all* its prime factors are upper numbers. As usual, this was finally proved by Euler, who used methods that may well have been the same as those of Fermat. But these methods must be left to Chapter 7, since they cannot be made clear until I have explained the idea of congruences.

6.

When the clock strikes thirteen

Solomon Grundy

Solomon Grundy
Born on Monday,
Christened on Tuesday,
Married on Wednesday,
Took sick on Thursday,
Worse on Friday
Died on Saturday,
Buried on Sunday;
This is the end of Solomon Grundy.

Suppose that on Monday morning I set out on a journey from which I return at the same time on Saturday. How long did the journey take?

Five days, you say? Was Solomon Grundy five days old when he died? How do you know it wasn't twelve days, or nineteen? I may have journeyed twenty years, like Odysseus, and finally reached home on a Saturday. There is no way to tell how many weeks I was away.

Now try another. I leave on Monday and return on Saturday, as before. On returning, I immediately rush out and make exactly the same journey again, taking the same amount of time. On what day of the week do I return?

Aha, you say. That problem is impossible. There isn't even enough information to fix the length of the first journey. Certainly the second journey is an unknown quantity.

But the problem is not impossible. The answer is, "Thursday." It is obvious, since I could start this journey on Monday and return on Saturday, that if I had started on Sunday I would have returned on Friday. This is true, even if the journey took twenty years instead of only five days. And in the same way, if I start the second time on Saturday, I must finish on Thursday (assuming that old age hasn't slowed me down).

As long as we are interested only in days of the week, it *doesn't matter* how many weeks went by during the journey. All that matters is how many extra days are added after a number of full weeks. A journey of 5 days (= 0 weeks and 5 days) and one of 12 days (= 1 week and 5 days) and one of 1083 days (= 154 weeks and 5 days) are all equivalent in that they will all end on a Saturday if begun on a Monday. Likewise they all will end on a Thursday if begun on a Saturday.

One way to look at this equivalence is that the numbers 5, 12, 1083 all give the same remainder (5) when divided by 7. Another way is that the difference between any two of them is a multiple of 7. Thus $1083 - 12 = 153 \times 7$. When we speak of such things, 7 is called a *modulus,* which means a measure, or a measuring rod. It is as though we used a rod to measure a distance, forgetting to count how many times we used the rod and taking note only of the fraction of the rod that was left over at the end. That is like forgetting how many weeks went by and taking note only of the day of the week. The length of the rod corresponds to 7, the length of the week in days.

There is also a special word to express the relationship between 5 and 12 or 1083. These numbers are all *congruent* to each other. Their congruence depends on 7 being the modulus; they would not be congruent if the modulus were 6. Therefore we say that they are congruent modulo 7, or

$$5 \equiv 12 \equiv 1083 \qquad (\text{mod } 7)$$

(The sign for congruence has three lines instead of two.)

If you have studied Latin, you will understand that "modulo 7" is an ablative absolute and means "7 being the modulus." In the eighteenth century, when congruence was first studied, most mathematical articles were written in Latin. The phrase, "modulo 7," was so catchy that it still sticks.

The Turtle Express

Let us try another problem. I receive two letters one day. One is postmarked Tuesday from Philadelphia, and the other is postmarked Monday from Chicago. Another day I receive two letters, and the one from Philadelphia is postmarked Thursday. What is the postmark on the one from Chicago?

The easiest way to handle this is by saying that letters must take a day longer to come from Chicago than from Philadelphia. Therefore the second letter from Chicago must be postmarked Wednesday, a day earlier than the second letter from Philadelphia.

But wait. I haven't said how the mail travels. Maybe it goes by air, maybe by bicycle, maybe by turtle. Perhaps that first letter from Chicago started off on Monday many weeks ago, and arrived together with one that left Philadelphia only last Tuesday.

Of course this doesn't change the answer. It only means that mail takes several weeks *and* a day longer to come from Chicago than from Philadelphia. If it comes from Chicago in C days, and from Philadelphia in P days, then

$$C - P \equiv 1 \pmod 7$$

It is still true that mail starting from Philadelphia on Thursday will reach me together with mail that started from Chicago, perhaps weeks ago, on Wednesday.

What if the trip from Chicago is *shorter* than the one from Philadelphia? The answer is still Wednesday. For example, suppose the trip from Chicago is six days shorter. Then a Monday letter from Chicago could arrive together with one sent the *previous* Tuesday from Philadelphia. But it still works out that a letter would have to leave Chicago on Wednesday, to arrive on the same day as a Philadelphia letter sent the previous Thursday.

In this case, $C - P = -6$. It still makes sense, though, to say that

$$C - P \equiv 1 \qquad (\text{mod } 7)$$

If you are careful you will see that the difference between -6 and 1 is just 7. Therefore -6 and 1 are congruent. To put it another way, the numbers 15, 8, 1, -6, -13 are all one *more* than a multiple of 7. (That is, -6 is one *more* than -7, just as 8 is one more than 7.) Therefore a negative number can be congruent to a positive number, and in problems about days of the week it makes no difference whether a number (such as $C - P$) is positive or negative, as long as we know what it is congruent to.

Congruent Arithmetic

Let us switch now, from days of the week to hours of the day. Here we have more to do with numbers, since the hours are numbered instead of being named. For example, if I start at two o'clock and work for three hours, it will be five o'clock when I finish, since $2 + 3 = 5$. But what if I start at ten and work for three hours? $10 + 3 = 13$. What time is it when the clock strikes thirteen?

All we need do is remember that the new modulus is 12, since there are twelve hours to the clock. Now, $13 \equiv 1$, mod 12. Therefore I shall finish my work at one o'clock.

Suppose I work from six o'clock to eight. Have I worked two hours? Not necessarily. I may have worked fourteen, since $14 \equiv 8 - 6$, mod 12. Again, if I sleep from eight to six, I may have had ten hours of sleep, or twenty-two, or even ninety-four, since $10 \equiv 22 \equiv 94 \equiv 6 - 8$, mod 12.

Now, the work I did from six to eight was so poor that I am obliged to do it over ten times, starting at eleven. When shall I finish?

Suppose that I can do the job once in x hours. We don't know whether $x = 2$ or 14 or something else, but we know that $x \equiv 2$, modulo 12. To do it ten times will take me $10x$ hours. Since $x \equiv 2$, modulo 12, it follows that $10x \equiv 20$, modulo 12. But I start at eleven and $11 + 20 = 31 \equiv 7$, modulo 12. Therefore I shall finish at seven o'clock.

Hold everything! We have cheated. We have taken the congruence

$$x \equiv 2$$

and multiplied both sides by 10 so as to get

$$10x \equiv 20$$

But multiplying both sides by 10 is something you are allowed to do only with an equation. You can't do it with a congruence—or can you?

Yes, you can do it with a congruence, and that is one reason why congruences are useful. For instead of writing $x \equiv 2$, modulo 12 we can write

$$x - 2 = \text{a multiple of } 12$$

This is an equation, so that we *can* multiply both sides by 10. Then we have

$$10x - 20 = 10 \text{ times a multiple of } 12$$

But 10 times a multiple of 12 is still a multiple of 12. Therefore

$$10x - 20 = \text{a multiple of } 12$$

which means that

$$10x \equiv 20 \qquad (\text{mod } 12)$$

To make sure of this, you may work out the problem step by step. I start work at eleven. When I have done the job once it will be one o'clock, perhaps three days later, but that doesn't matter. When I have done it twice it will be three o'clock, again maybe three days later. The third time I shall finish at five, the fourth time at seven, the fifth at nine. Just keep counting and you will find that the tenth job will be done at seven.

Congruences behave very much like equations. You can add and subtract them as well as multiply. Let us say that I always sit down to work at six, and sometimes I do my homework, which takes until eight, and sometimes I write letters, which takes until seven (but we don't say how many days later). One day I have both homework *and* letters. What time do I finish?

The easy way is the right way. If homework takes x hours and letters take y hours, then

$$x \equiv 2 \qquad (\text{mod } 12)$$
$$y \equiv 1 \qquad (\text{mod } 12)$$

Adding, we find

$$x + y \equiv 3 \qquad (\text{mod } 12)$$

and therefore, starting at six o'clock, I finish at nine.

Here we have added two congruences together. This is all right, for if we wrote the congruences as equations, we should see that $x - 2$ and $y - 1$ are both multiples of 12. But the sum of two multiples of 12 is certainly a multiple of 12. (Thus, $24 + 36 = 60$.) The sum of $x - 2$ and $y - 1$ is $x + y - 3$. Therefore $x + y - 3$ is a multiple of 12, or

$$x + y \equiv 3 \qquad (\text{mod } 12)$$

It is easier and shorter, though, just to add the congruences.

The difference of two multiples of 12 is also a multiple of 12.

Therefore congruences can be subtracted as well as added.

When we multiplied both sides of a congruence by 10, we actually were multiplying a congruence

$$x \equiv 2 \pmod{12}$$

by an equation

$$10 = 10$$

But it is also possible to multiply two congruences together. Suppose that

$$a \equiv b \pmod{m}$$
$$c \equiv d \pmod{m}$$

We can surely multiply the first congruence by c and the second by b. This gives

$$ac \equiv bc \quad \text{and} \quad bc \equiv bd$$

and therefore

$$ac \equiv bd \pmod{m}$$

which is just what we get by multiplying together the original two congruences.

Note that in this proof we have used another rule that applies to equations. We have assumed that if two numbers (ac and bd) are congruent to a third number (bc) then they are congruent to each other. Or generally, if $x \equiv y$ and $y \equiv z$, then $x \equiv z$. Try and prove this yourself. You win if you can show that $x - z$ is a multiple of m, supposing that $x - y$ and $y - z$ are multiples of m.

Moby Dick

Here is a problem in multiplying congruences. Suppose that

Captain Ahab can catch and kill a whale in an hour. But each time he kills one he has to stomp up and down the deck for a certain number of hours, cursing the whale for not being Moby Dick. If he starts cursing a whale at three o'clock, he finishes at two. One day he sets sail at nine o'clock, and hunts whales without stopping until he returns at four o'clock. He saved up all his stomping till after dinner. If he begins stomping at eight o'clock, what time will it be when he stops? (He has to stomp separately for each whale that he killed, because he can curse only one whale at a time.)

We don't know how many *days* he spent hunting whales, and we don't know how many *days* he stomps for each whale. But neither of these facts is needed. Suppose he hunted for h hours and stomps for s hours per whale. Then we do know that

$$h \equiv 4 - 9 \equiv 16 - 9 = 7 \qquad \pmod{12}$$
$$s \equiv 2 - 3 \equiv 14 - 3 = 11 \qquad \pmod{12}$$

Since he catches whales at the rate of one an hour, he has h whales to stomp for. This will take sh hours. But

$$sh \equiv 11 \text{ times } 7 = 77 \equiv 5 \qquad \pmod{12}$$

Since he begins at eight, and

$$8 + 5 \equiv 1 \qquad \pmod{12}$$

he finishes stomping at one o'clock, just in time for lunch.*

By the way, do you remember my homework? I began it at six and finished at eight. I suppose you could tell me easily what time it was when I had done half of it?

Beware! It may have been seven o'clock, but not necessarily. If I started at 6 P.M. and finished the next morning, then I was half done at 1 A.M.!

This is puzzling. Can't it be reasoned out by congruences? Let's see, if half the job takes z hours, then the whole job takes $2z$ hours. Therefore

* See note in Appendix 1.

$$2z \equiv 2 \qquad (\text{mod } 12)$$

Doesn't that mean that $z \equiv 1$?

No, it doesn't. Clearly, z might be 7, which is *not* congruent to 1, since $7 - 1$ is not a multiple of 12. But then $2z$ would be 14, which *is* congruent to 2. In other words, we are *not allowed to divide congruences by anything.* But for adding, subtracting, and multiplying, congruences can be treated just as if they were equations.

The law against dividing breaks down when the modulus is a prime number. For suppose m is prime, and

$$xz \equiv yz \qquad (\text{mod } m)$$

Then $xz - yz$, or $(x - y)z$, is divisible by m. But that means that either $x - y$ or z is divisible by m, by the unique factorization theorem of Chapter 3. In other words, either $z \equiv 0$, or $x \equiv y$. Thus if the modulus is prime *and z is not congruent to zero,* we may divide the original congruence by z and obtain

$$x \equiv y \qquad (\text{mod } m)$$

Congruences with a prime modulus behave just like ordinary equations, which also can be divided by anything except zero.

Suppose, for example, that I start a journey at noon on Wednesday and finish it at noon on an unknown day. Immediately I repeat the journey and finish it the second time at noon on Thursday. On what day did I finish the first time?

Say that the journey took z days (z is known to be a whole number, since I finished at noon). Then

$$2z \equiv \text{Thursday minus Wednesday} \equiv 1 \qquad (\text{mod } 7)$$

We must not deduce that $z = ½$, since z is not a fraction. Instead we can write

$$2z \equiv 1 \equiv 8 \qquad (\text{mod } 7)$$

and therefore

$$z \equiv 4 \qquad (\text{mod } 7)$$

Therefore the journey took from Wednesday to Sunday (any Sunday) and again from Sunday to Thursday.

Note that I have divided a congruence by 2, which is allowed since 7 is prime. This is the operation that led to a wrong answer in the clock problem, where the modulus, 12, was not prime.

The division would also be forbidden if the divisor (2 in this case) were a multiple of 7, for such a divisor would be congruent to zero. Thus, if the journey were made fourteen times, the problem could not be solved. For then the last journey would be surely finished on Wednesday, the starting day, no matter what day the first journey ended.

Casting Out Nines

Congruences are easy to understand when the modulus is 10. If a number is divided by 10, the remainder will be the last digit of the number. So two numbers are congruent mod 10 if they have the same last digit, and not otherwise. This makes it easy to tell the last digit of a sum. For instances,

$$79742 \equiv 2 \qquad (\text{mod } 10)$$
$$6134 \equiv 4 \qquad (\text{mod } 10)$$

Adding two congruences, we have

$$79742 + 6134 \equiv 2 + 4 \equiv 6 \qquad (\text{mod } 10)$$

Therefore the sum of 79742 and 6134 must end in 6. This is no surprise, because the first thing you would do in adding them together is to write down the last digit of the sum

$$\begin{array}{r} 79742 \\ 6134 \\ \hline 6 \end{array}$$

In the same way, we can see that

$$79742 - 6134 \equiv 2 - 4 \equiv 8 \pmod{10}$$

and

$$79742 \times 6134 \equiv 2 \times 4 \equiv 8 \pmod{10}$$

which is just what you would find when you began the subtraction or multiplication.

We *cannot* so easily find the last digit of $79742 \div 6134$. Remember that congruences cannot be divided. Now $79742 \div 6134$ *may* be congruent to 3 ($\equiv 12 \div 4$) or it *may* be congruent to 8 ($\equiv 32 \div 4$). The only way to find out is by long division, which shows that $79742 \div 6134 = 13 \equiv 3$. Division is the only operation of arithmetic in which you start from the first instead of the last digit. The real reason for this is that in division the last digit of the answer cannot be decided by congruences, but depends on all the other digits.

A very useful modulus is 9, because it makes all the powers of 10 congruent to 1. Thus

$$10 \equiv 1 \pmod{9}$$
$$100 = 99 + 1 \equiv 1 \pmod{9}$$
$$1000000 = 999999 + 1 \equiv 1 \pmod{9}$$

and so on. Another way to see this is by multiplying congruences

$$\begin{aligned} 1000000 &= 10 \times 10 \times 10 \times 10 \times 10 \times 10 \\ &\equiv 1 \times 1 \times 1 \times 1 \times 1 \times 1 = 1 \end{aligned}$$

From this it follows that

$$500 = 100 \times 5 \equiv 1 \times 5 = 5 \pmod{9}$$
$$40 \equiv 4 \pmod{9}$$
$$6000 \equiv 6 \pmod{9}$$

and so on. Therefore

$$\begin{aligned} 6548 &= 6000 + 500 + 40 + 8 \\ &\equiv 6 + 5 + 4 + 8 = 23 \equiv 5 \qquad \pmod{9} \end{aligned}$$

You may check this by dividing 6548 by 9 and getting a remainder of 5.

In other words, any number is congruent, modulo 9, to the sum of its digits. For example

$$\begin{aligned} 8327769 &\equiv 8 + 3 + 2 + 7 + 7 + 6 + 9 \\ &= 42 \\ &\equiv 4 + 2 \\ &= 6 \end{aligned}$$

This work is much easier than dividing 8327769 by 9 to obtain a remainder of 6. It can be made even easier by "casting out nines" as we go. Since 9 is congruent to 0, the digit 9 can be left out of the sum. Also any group that adds up to 9 can be left out. Thus,

$$\begin{aligned} 8327769 &\equiv 8 + 3 + [2 + 7] + 7 + 6 + [9] \\ &\equiv 8 + 3 + 7 + 6 = 24 \\ &\equiv 2 + 4 = 6 \end{aligned}$$

The method of casting out nines is useful for checking long arithmetic problems. Thus

$$\begin{array}{rll} 718 & \equiv 7 + \cancel{1 + 8} \equiv & 7 \\ \times 364 & \equiv \cancel{3 + 6} + 4 \equiv & \underline{4} \\ \hline 2872 & & 28 \equiv 2 + 8 = 10 \equiv 1 \\ 4308 & & \\ \underline{2154} & & \\ 261352 & \equiv \cancel{2} + \cancel{6} + \cancel{1} + 3 + 5 + 2 \equiv 10 \equiv 1 & \end{array}$$

The two factors are congruent to 7 and 4, so that their product must be congruent to 28. And 261352 *is* congruent to 28 (that is, to 1). This doesn't guarantee that it is right, but if it came out not congruent to 28 it would surely be wrong.

Another example:

$$\begin{array}{rl} 721096 \equiv \cancel{7} + \cancel{2} + 1 + 6 & \equiv 7 \\ -\underline{487682} \equiv 4 + 8 + 7 + 6 + 8 + 2 = 35 & \equiv \underline{8} \\ 233414 \equiv 2 + 3 + 3 + \cancel{4} + \cancel{1} + \cancel{4} \equiv 8 & \equiv -1 \end{array}$$

The subtraction is congruent to $7 - 8$, which is -1. Sure enough, 233414 is congruent to 8, which is congruent to -1.

Casting out nines also shows whether a number is a multiple of 9. All multiples of 9 are congruent to each other, so that if we add up the digits of 71352 and get 18, which is a multiple of 9, then we know that 9 goes into 71352.

All this works for 3 exactly as it does for 9. The reason is that $10 \equiv 1$, modulo 3, and everything follows from that. Arithmetic can be checked by casting out threes, but that is not as good as casting out nines because with threes there is too much chance that the congruences may check by accident even though the answer is really wrong. However, adding up digits is a good way of finding whether 3 goes into a number.

Eleven is an interesting modulus, because

$$10 \equiv -1 \pmod{11}$$

Multiplying 10 by itself, we have

$$\begin{aligned} 100 &\equiv (-1) \times (-1) \equiv 1 \pmod{11} \\ 1000 &\equiv (-1) \times (-1) \times (-1) \equiv -1 \pmod{11} \\ 10000 &\equiv 1 \pmod{11} \end{aligned}$$

and so on. (You can check these statements by dividing by 11. For example, 11 goes 91 times into 1000, with a remainder of -1.)

This means that in casting out elevens the digits must be alternately added and subtracted, starting from the right. Thus

$$\begin{aligned} 7256 &= 6 + 50 + 200 + 7000 \\ &\equiv 6 - 5 + 2 - 7 = -4 \\ &\equiv 7 \qquad \pmod{11} \end{aligned}$$

Therefore 7256 is not divisible by 11. But

$$7656 \equiv 6 - 5 + 6 - 7 = 0$$

so that 7656 is a multiple of 11. These tricks for checking multiples of 3 and 11 were mentioned in Chapter 3.

For most moduli, such as 23, 10 is not congruent to 1 or -1, and there are no simple tricks with digits. Congruences are still useful, though, in finding whether a small number like 23 goes into a big number that we do not want to calculate. For example, from Chapter 3 we know that $2^{10}(2^{11}-1)$ will be a perfect number if $2^{11}-1$ is prime. But $2^{11}-1$ is not prime; it is divisible by 23, and this can be shown by congruences without even having to find the value of 2^{11}. Simply write

$$2^5 = 32 \equiv 9 \qquad \pmod{23}$$

and square both sides

$$2^{10} \equiv 9^2 = 81 \equiv 12 \qquad \pmod{23}$$

Multiply by 2

$$2^{11} \equiv 2 \times 12 = 24 \equiv 1 \qquad \pmod{23}$$

Therefore $2^{11}-1$ is a multiple of 23.

Congruences also make it possible to prove some theorems just by checking a few numbers. A theorem mentioned in Chapter 1 is that the square of an odd number is always 1 more than a multiple of 8. This may be written,

$$\text{if } x \text{ is odd, then } x^2 \equiv 1 \qquad \pmod 8$$

To prove this, note that every odd number is congruent to 1, 3, 5, or 7, modulo 8. Therefore

$$\text{if } 1^2 \equiv 3^2 \equiv 5^2 \equiv 7^2 \equiv 1 \pmod 8$$

then the theorem will be true for any odd x! Just check these four squares ($1^2 = 0 + 1$, $3^2 = 8 + 1$, $5^2 = 24 + 1$, $7^2 = 48 + 1$) and you have proved the theorem.

A similar theorem is that every cube is either a multiple of 9 or is right next to one. (Thus, $3^3 = 27 = 9 \times 3$, $5^3 = 125 = 9 \times 14 - 1$, $7^3 = 343 = 9 \times 38 + 1$.) Just check it for all the cubes from 0^3 to 8^3, and it is proved for all numbers. In fact, you need only check up to 4^3. If it is true of 4^3, then it is obviously true of $(-4)^3$. But $5 \equiv -4$, mod 9, and therefore it is true of 5^3. Similarly, $6 \equiv -3$, $7 \equiv -2$, $8 \equiv -1$.

In the course of checking the last theorem, you may stumble on the fact that every number is congruent to its own cube, modulo 3. That is,

$$x^3 \equiv x \pmod 3 \qquad \text{for all } x$$

This is almost obvious, since $1^3 = 1$, $0^3 = 0$, $(-1)^3 = -1$, and every number is congruent, modulo 3, to 1, 0, or -1. It may remind you of an easier theorem,

$$x^2 \equiv x \pmod 2 \qquad \text{for all } x$$

This one just says that the square of an odd number is odd, of an even number, even.

We are tempted now to guess that

$$x^4 \equiv x \pmod 4$$

Alas! Let $x = 2$, then $x^4 = 16$, which is not congruent to 2, modulo 4.

It would be natural to stop after this failure, but Pierre Fermat didn't. He tried 5 as modulus. Every number is congruent to 0, 1, -1, 2, or -2, modulo 5. If $x = 0$, 1, or -1, it is obvious that $x^5 = x$. Therefore we need only work out $x = 2$. (The case of -2 will work out the same.) We have

$$2^5 = 32 \equiv 2 \qquad (\text{mod } 5)$$

Therefore

$$x^5 \equiv x \qquad (\text{mod } 5) \qquad \text{for all } x$$

You may see for yourself that this doesn't work if the modulus is 6 (try $x = 2$), but that it does work if the modulus is 7 (you need only check 2^7 and 3^7).

Fermat announced that the rule works for any prime modulus. That is,

$$x^p \equiv x \qquad (\text{mod } p) \qquad \text{for all } x \text{ and all } \textit{prime } p$$

This lovely fact is called Fermat's Theorem. Don't confuse it with Fermat's Last Theorem, which came up in Chapter 5. The two theorems are not related, although their names are similar, and mathematicians never say one when they mean the other. Fermat's Last Theorem has been called a theorem only by courtesy since the proof was lacking until claimed in 1993. But Fermat's theorem has all along been solid; it was proved by Euler, in two different ways. (See Chapter 8.)

Fermat's Theorem cannot be proved by checking a few numbers. That will do for any *single* value of p, since any x is congruent mod p to a number $\leq p$. But the checking must be done over for each *new* p, since congruences do not tell us how to move from one modulus to another. It would take forever to check all prime values of p. The theorem must be proved some other way. I shall explain Euler's proofs in Chapter 8. For now, I shall mention some clever uses to which Euler put this theorem. Bear in mind that if x is not congruent to zero, we can divide by it and get $x^{p-1} \equiv 1$, modulo p, for all prime p and all x not divisible by p. This statement also is called Fermat's Theorem.

How to Save Sieving

You remember that Fermat had claimed to know exactly which numbers are primitive hypotenuses (sums of two squares with no common factor). Euler was fascinated by this problem and worked at it for many years, off and on. In Chapter 5, I spoke of upper numbers and lower numbers. Upper numbers are congruent to 1, modulo 4, and lower numbers are congruent to -1, that is, to 3. Now, it is easy to show that a lower number cannot be a sum $x^2 + y^2$. All odd squares are congruent to 1, modulo 4, and all even squares are congruent to 0. Therefore the sum of two squares is always congruent to 2, 1, or 0, never to 3. But Fermat went further and claimed that no primitive hypotenuse is even *divisible* by any lower number. This Euler was able to prove, with the aid of Fermat's Theorem.

Any lower number has a lower prime factor, since the product of any number of upper primes would itself be upper. Therefore it is enough to show that a primitive hypotenuse has no lower *prime* factor. Now, if p is a lower prime, then $p = 2n + 1$, where n is odd. Suppose that $x^2 + y^2$ is a multiple of p, but x and y have no common factor. Then

$$x^2 \equiv -y^2 \pmod{p}$$

and if we take the nth power of both sides, then

$$x^{2n} \equiv -y^{2n} \pmod{p}$$

Euler pointed out that this is impossible since $2n = p - 1$, and Fermat's Theorem says that $x^{p-1} \equiv 1 \equiv y^{p-1}$ modulo p.

If p is an upper prime, then n is even, and we get

$$x^{2n} \equiv y^{2n} \qquad (\text{mod } p)$$

which does not contradict Fermat's Theorem.

Euler applied these ideas also to the problem of finding perfect numbers. It depends on knowing whether $2^n - 1$ is prime. Now, $2^n - 1$ can't be prime if n is not prime. For if $n = ab$, let $q = 2^a - 1$. Then

$$2^a \equiv 1 \qquad (\text{mod } q)$$

and raising both sides to the power b, we have

$$2^{ab} \equiv 1^b \equiv 1$$

or

$$2^n \equiv 1 \qquad (\text{mod } q)$$

Therefore $2^n - 1$ has the factor q.

The hard question is, supposing that n *is* prime, when will $2^n - 1$ *not* be prime? That is, when is there another prime, p, such that

$$2^n \equiv 1 \qquad (\text{mod } p)$$

Euler made the surprising discovery that this is impossible unless

$$p \equiv 1 \qquad (\text{mod } n)$$

He started with Fermat's Theorem, saying that

$$2^{p-1} \equiv 1 \qquad (\text{mod } p)$$

Let f be the greatest common factor of n and $p - 1$. Then by the GCF theorem of Chapter 3, we have $f = cn + d(p - 1)$ or

$$\begin{aligned} 2^f &= 2^{cn} \times 2^{d(p-1)} \\ &= (2^n)^c \times (2^{p-1})^d \end{aligned}$$

If we assume that

$$2^n \equiv 1 \pmod{p}$$

then we have

$$\begin{aligned} 2^f &\equiv 1^c \times 1^d \\ &\equiv 1 \pmod{p} \end{aligned}$$

On the other hand, n is supposed to be prime. Therefore f, being a factor of n, must be either 1 or n. If $f = 1$ then $2^f = 2$, and

$$2 \equiv 1 \pmod{p}$$

which is impossible. So $f = n$, which means that n is a factor of $p - 1$, or

$$p \equiv 1 \pmod{n}$$

This result is no surprise when $2^n - 1$ is prime, since it just says that $2^n - 1 \equiv 1$ or $2^n \equiv 2$, modulo n, which is Fermat's Theorem. But now we have proved that every prime factor (and hence *every* factor) of $2^n - 1$ is also congruent to 1. For example, $2^{11} - 1 = 2047 = 23 \times 89$. Both 23 and 89 are 1 more than a multiple of 11.

This helps a great deal in finding new perfects. If we wished to test $2^{13} - 1$ by the sieve of Eratosthenes, we should have to divide it by every prime up to about 90, which is approximately the square root of $2^{13} - 1$. But now we can ignore any prime which is not 1 more than a multiple of 13. The only primes below 100 that we need consider are 53 and 79. With 53 as modulus

$$\begin{aligned} 2^6 &= 64 \\ &\equiv 11 \\ 2^{12} &\equiv 121 \equiv 15 \\ 2^{13} - 1 &\equiv 29 \qquad \text{(not congruent to zero)} \end{aligned}$$

and with 79 as modulus

$$\begin{aligned} 2^6 &= 64 \\ &\equiv -15 \\ 2^{12} &\equiv 225 \equiv -12 \\ 2^{13} - 1 &\equiv -25 \end{aligned}$$

so that neither 53 nor 79 goes exactly into $2^{13} - 1$. Therefore $2^{13} - 1$ is prime, and $2^{12}(2^{13} - 1)$ is perfect. (It is the perfect number we used for an "aliquot will" in Chapter 3.)

Euler used the same trick to find factors of Fermat numbers, which you remember are numbers like $2^n + 1$ where n is itself a power of 2. If $2^n + 1$ has a prime factor p, then

$$2^n \equiv -1 \qquad \text{and} \qquad 2^{2n} \equiv 1 \qquad (\text{mod } p)$$

By the GCF theorem, $p - 1$ and $2n$ must have a common factor f such that

$$2^f \equiv 1 \qquad (\text{mod } p)$$

Now f can't be a factor of n since then 2^n would also be congruent to 1, instead of to -1. But *since n is a power of* 2, the only number that goes into $2n$ and not into n is $2n$ itself. Therefore $f = 2n$ and $p - 1$ is divisible by $2n$, or

$$p \equiv 1 \qquad (\text{mod } 2n)$$

This is such a strong condition that it is not surprising that the first few Fermat numbers are prime. Take $n = 8$, $2^n + 1 = 257$. Any factor of 257 must be congruent to 1, modulo 16. But the smallest such number is 17, which is already too big since $17^2 = 289$. Therefore 257 is prime.

Next let $n = 16$. In Chapter 3, I pointed out that to check $2^{16} + 1$ by the sieve of Eratosthenes would take hours of hard work since every prime up to 250 must be tested. But now we only need consider primes congruent to 1, modulo 32. There are two of these below 256: 97 and 193.

modulo 97: $2^5 \times 3 \equiv -1$

(cube both sides) $2^{15} \times 27 \equiv -1$

(multiply by 2) $2^{16} \times 27 \equiv -2$

which is impossible if $2^{16} \equiv -1$.

modulo 193: $2^6 \times 3 \equiv -1$

(square) $2^{12} \times 9 \equiv 1$

(multiply by 2^4) $2^{16} \times 9 \equiv 16$

which is impossible if $2^{16} \equiv -1$.

Therefore $2^{16} + 1$ is not divisible by 97 or 193, and must be prime. The work of hours has been reduced to a few lines.

Euler went on to attack $2^{32} + 1$. The possible prime factors are congruent to 1, modulo 64. This narrows the field a lot, but now we may have to test primes up to 65000 or so. However, Euler started out boldly, hoping to find a factor early.

modulo 193:

$$2^6 \times 3 \equiv -1$$
$$2^{30} \times 3^5 \equiv -1$$
$$2^{32} \times 3^5 \equiv -4$$

but $3^5 = 243 \equiv 50$, therefore 2^{32} cannot $\equiv -1$.

modulo 257:

$$2^8 \equiv -1$$
$$2^{32} \equiv 1, \qquad \text{not } -1$$

modulo 449:

$$2^6 \times 7 \equiv -1$$
$$2^{30} \times 7^5 \equiv -1$$
$$2^{32} \times 7^5 \equiv -4, \qquad \text{but } 7^5 = 343 \times 49 \equiv -106 \times 49$$
$$= -5194 \equiv -255$$

therefore 2^{32} cannot $\equiv -1$.

modulo 577:

$$2^6 \times 9 \equiv -1$$
$$2^{30} \times 9^5 \equiv -1$$
$$2^{32} \times 9^5 \equiv -4, \qquad \text{but } 9^5 = 729 \times 81 \equiv 152 \times 81$$
$$= 1368 \times 9 \equiv 214 \times 9 = 1926 \equiv 195$$

therefore 2^{32} cannot $\equiv -1$.

modulo 641:

$$2^7 \times 5 \equiv -1$$
$$2^{28} \times 5^4 \equiv 1$$
$$2^{32} \times 5^4 \equiv 16, \qquad \text{but } 5^4 = 625 \equiv -16$$

therefore $2^{32} \equiv -1$, or $2^{32} + 1$ is divisible by 641.

And this is how Euler discovered that $2^{32} + 1$ is not prime. Evidently Fermat never thought of this method, although all the ingredients were known to him. The discovery was one of Euler's earliest, made before he was twenty-five. At that time (1732), he had not yet proved Fermat's Theorem, but merely assumed it! No one, though, could deny that 641 was a factor of $2^{32} + 1$, once Euler had pointed it out.

7.

Hard nuts

Factoring a Hypotenuse

The fact that a primitive hypotenuse cannot be divisible by a lower prime can be proved without Fermat's Theorem, by a different method which is worth studying because it leads to the solution of harder problems as well.

First I remark that if $x^2 + y^2$ is a multiple of p and x and y have no common factor, then many other multiples of p can be found which are the sum of two squares. This is done by congruences. We have

$$x^2 + y^2 \equiv 0 \pmod{p}$$

and therefore if $a \equiv x$ and $b \equiv y$, modulo p, then

$$a^2 + b^2 \equiv 0 \pmod{p}$$

Obviously there are many ways to choose a and b. If we like, we may choose them both to be less than p, since there is always some number, less than the modulus, which is congruent to any given x or y. But we can do even better, and find both a and b less than *half* of p. This is because if a is greater, then $p - a$ is less than half of p. But $p - a$ is congruent to $-a$, and so its square is congruent to the square of $-a$, which is a^2. Therefore if $a^2 + b^2$ is a multiple of p, then so is $(p - a)^2 + b^2$. The same goes for b and $p - b$.

The upshot is that if *some* multiple of p is a primitive hypotenuse, then there are two numbers a and b, both less than $\frac{1}{2}p$, such that $a^2 + b^2$ is a multiple of p. (Note that neither x nor y can be divisible by p, since then both would be, and we know they have no common factor. Therefore neither a nor b is zero.)

Since the game is to make $a^2 + b^2$ as small as possible, we observe that if $a^2 + b^2$ is even, a and b must be either both odd or both even. Thus we can find u and v so that

$$a = u + v$$
$$b = u - v$$

and therefore

$$a^2 + b^2 = 2(u^2 + v^2)$$

(Check this yourself.) Since p is supposed to be an odd prime, it will go into $u^2 + v^2$ if it goes into $a^2 + b^2$. Again, if $u^2 + v^2$ is even, then we can find t and s so that

$$u^2 + v^2 = 2(t^2 + s^2)$$

and p goes into $t^2 + s^2$. This goes on until we arrive at an *odd* number $c^2 + d^2$ which is a multiple of p.

Now $a^2 + b^2$ may have been odd to begin with; then $c = a$ and $d = b$. But in any case $c^2 + d^2$ is no bigger than $a^2 + b^2$, which is less than p^2 since a and b are less than $\frac{1}{2}p$. Let us write

$$c^2 + d^2 = pq$$

Then q is odd and less than p, since $c^2 + d^2$ is odd and less than p^2. Therefore, if p is prime and goes into some primitive hypotenuse $x^2 + y^2$, then there is an odd number q, less than p, such that pq is a hypotenuse. In fact, we can make pq a *primitive* hypotenuse since if c and d have a common factor, the factor can be removed, and the new $c^2 + d^2$ will be even smaller and still odd and divisible by p. (The common factor cannot be divisible by p since c and d are less than $\frac{1}{2}p$ to begin with.)

Let us see whether this is possible if $p = 3$. Then q would have to be 1. But $3 \times 1 = 3$ which cannot *itself* be a hypotenuse, since it is a lower number (congruent to -1, modulo 4). Therefore 3 cannot be a factor of any primitive hypotenuse.

Now suppose $p = 7$. Then q must be 1, 3, or 5. Both 1×7 and 5×7 are lower numbers and cannot be the sum of two squares. And 3×7 is not a primitive hypotenuse because it has the factor 3. Therefore no multiple of 7 is a primitive hypotenuse.

We could go on like this, taking all the lower primes in order. If p is a lower prime, q may be either an upper or a lower number. But if q is upper, then pq is lower and cannot be the sum of two squares. And if q is lower, it has at least one lower prime factor r, which we have already considered before we considered p. Therefore pq, which is a multiple of r, is not a primitive hypotenuse.

In this way we prove that no primitive hypotenuse has a lower prime factor. This is a proof by induction, and I should like to present it again in the form of an infinite descent. The proof has several parts, and I shall draw your attention to some of the parts by calling them lemmas. A lemma is a theorem that is stated as part of the proof of another theorem.

Lemma 1. If any primitive hypotenuse has a prime factor p, then p is also a factor of some primitive hypotenuse less than p^2.

This is just what I proved by using congruences on pp. 136–137.

Lemma 2. A sum $a^2 + b^2$ cannot have just one lower prime factor. It has either none or at least two.

By congruences modulo 4, it is easy to see that the product of any number of upper primes with one lower prime is a lower number, and cannot be a sum $a^2 + b^2$. This proves Lemma 2 for odd sums $a^2 + b^2$. But an even sum $a^2 + b^2$ is always (see pp. 136–137) double another such sum, which has the same number of lower prime factors as the original sum $a^2 + b^2$. The new $c^2 + d^2$, too, is either odd or double another, and so on until we arrive at an odd sum $x^2 + y^2$, for which the lemma has already been proved. Therefore Lemma 2 is completely true.

Once these two lemmas are accepted, the infinite descent is quite simple. Suppose that h is a primitive hypotenuse and has a lower prime factor p. Then by Lemma 2, h has a second lower prime factor q. The product pq is either equal to h or a factor of h. Anyway, it isn't more than h. Now, pq lies between p^2 and q^2, so that p^2 and q^2 can't *both* be bigger than h. Suppose p^2 is not bigger than h. By Lemma 1, there is a primitive hypotenuse h', which is divisible by p and less than p^2. Hence h' is less than h, and we start over with h' in place of h. (If p^2 is bigger than h, and q^2 is not, then we use q in place of p.) In this way we find an infinite descending sequence (h, h', etc.) of primitive hypotenuses, each with a lower prime factor. This is impossible.

The nice thing about this proof is that you never have to worry about what a lower prime actually is, except when you are proving Lemma 2. This means that the same proof works for any class of prime factors about which you can prove the equivalent of Lemma 2. For example, let us call a prime number "good" if it is the sum of two squares, and "bad" if it is not. Euler found a way to prove the following statement.

Lemma 2a. A primitive hypotenuse cannot have just one bad prime factor. It has either none or at least two.

Once Euler had shown this, he used exactly the same argument by infinite descent to prove the important theorem that a primitive hypotenuse has *no* bad prime factor. In other words, every prime factor of a primitive hypotenuse is itself a primitive hypotenuse. I shall call this the factor theorem for primitive hypotenuses.

There remained only one step before the whole problem was

solved. This was to show that every upper prime has some multiple which is a primitive hypotenuse. Euler finally proved this, by a method I leave to Chapter 9. Then, by the factor theorem, he knew that every upper prime is itself a primitive hypotenuse. This is just the first of Fermat's two statements about primitive hypotenuses (end of Chapter 5); the second he had already proved. Thus Euler conquered the whole subject.

I now return to Euler's proof of Lemma 2a. As I mentioned in Chapter 4, it is easy to make bigger sums $a^2 + b^2$ by multiplying smaller ones together. But Euler noticed that this process sometimes can lead to *smaller* such sums. For example,

$$5 = 2^2 + 1^2$$
$$65 = 7^2 + 4^2$$

and therefore

$$5 \times 65 = (2 \times 7 + 1 \times 4)^2 + (2 \times 4 - 1 \times 7)^2$$

or $$325 = 18^2 + 1^2$$

so that 325 is a primitive hypotenuse. But we can also write

$$5 \times 65 = (2 \times 7 - 1 \times 4)^2 + (2 \times 4 + 1 \times 7)^2$$
$$325 = 10^2 + 15^2$$

Now, 10 and 15 both have the factor 5. Therefore the whole equation can be divided by 25

$$13 = 2^2 + 3^2$$

So, starting with 5 and 65, we end up with 13.

It is no accident that 10 and 15 have the factor 5 in common. It comes about because 5 and 65 both have this factor. We may express this in a theorem.

If p is prime and

$$p = a^2 + b^2$$
$$pq = x^2 + y^2$$

then either $ax + by$ and $ay - bx$ both have the factor p, or else $ax - by$ and $ay + bx$ both have the factor p.

This is proved by congruences, modulo p.

$$a^2 \equiv -b^2$$
$$x^2 \equiv -y^2$$

and therefore

$$a^2x^2 \equiv b^2y^2$$

But $a^2x^2 - b^2y^2 = (ax + by)(ax - by)$. Hence, by the unique factorization theorem, either $ax + by$ or $ax - by$ is a multiple of p.

Suppose $ax + by$ is a multiple of p. Then

$$\begin{aligned} ax &\equiv -by \\ abx &\equiv -b^2y \\ &\equiv a^2y \qquad (\text{since } a^2 \equiv -b^2) \end{aligned}$$

and therefore $a(ay - bx)$ is divisible by p. But a is less than p; therefore p cannot go into a and must go into $ay - bx$.

In the same way, if $ax - by$ is a multiple of p, then so is $ay + bx$. This proves the theorem.

It follows that q is a hypotenuse. For if

$$ax + by = pu$$

and

$$ay - bx = pv$$

then

$$\begin{aligned} (pu)^2 + (pv)^2 &= (a^2 + b^2) \times (x^2 + y^2) \\ &= p^2q \end{aligned}$$

Dividing by p^2, we have

$$u^2 + v^2 = q$$

Now we can prove Lemma 2a. Suppose that h is a primitive hypotenuse with just one bad prime factor r. Let p be a good prime factor of h, so that $h = pq$. Then the theorem we just proved shows that $q = u^2 + v^2$. Now if q has a good prime factor p', then $q = p'q'$ and the same theorem shows that $q' = u'^2 + v'^2$. In this way we get a descending sequence of hypotenuses h, q, q', etc., from which we remove one good prime factor after another. (These hypotenuses may not be primitive, but that does not spoil the proof.) Since r is the only bad prime factor, all the other factors can be removed and we end by showing that r is a sum $x^2 + y^2$. But then r is not a bad prime after all. This proves Lemma 2a, and the factor theorem follows automatically.

Variations on a Theme

After this tough struggle with hypotenuses, you may feel dismayed when I bring up a more difficult question: what numbers are equal to the sum of a square and *double* another square?

Take heart. We shall solve this very quickly by exactly the same methods that worked for hypotenuses. First, the multiplication theorem. Suppose

$$g = a^2 + 2b^2$$

and

$$h = c^2 + 2d^2$$

Then

$$\begin{aligned} gh &= a^2c^2 + 2a^2d^2 + 2b^2c^2 + 4b^2d^2 \\ &= (a^2c^2 + 4abcd + 4b^2d^2) + (2a^2d^2 - 4abcd + 2b^2c^2) \\ &= (ac + 2bd)^2 + 2(ad - bc)^2 \end{aligned}$$

For example, from

$$9 = 1^2 + 2 \times 2^2$$
$$11 = 3^2 + 2 \times 1^2$$

we get

$$99 = (1 \times 3 + 2 \times 2 \times 1)^2 + 2(3 \times 2 - 1 \times 1)^2$$
$$= 7^2 + 2 \times 5^2$$

Next, Lemma 2. Let us call a prime "2-good," if it is of the form $a^2 + 2b^2$. (Two itself is 2-good, let $a = 0$, $b = 1$. Also $3 = 1^2 + 2 \times 1^2$ is 2-good, but 5 is not of this form and is therefore 2-bad.) Then we may prove Lemma 2b.

Lemma 2b. If x and y have no common factor, then $x^2 + 2y^2$ cannot have just one 2-bad prime factor. It has either none or at least two.

This is proved in exactly the same way as Lemma 2a.

If $$p = a^2 + 2b^2$$

and $$pq = x^2 + 2y^2$$

then, modulo p,

$$a^2 \equiv -2b^2$$
$$x^2 \equiv -2y^2$$
$$a^2x^2 \equiv 4b^2y^2$$

and hence, if p is prime, either $ax + 2by$ or $ax - 2by$ is a multiple of p, say pu. Further, either $ay - bx$ or $ay + bx$ is a multiple of p, say pv. (Do this part yourself.) Either way

$$p^2q = (pu)^2 + 2(pv)^2$$

or $$q = u^2 + 2v^2$$

Now if $q = p'q'$, where p' is 2-good, then

$$q' = u'^2 + 2v'^2$$

by the same reasoning. Finally, we have removed all the 2-good prime factors of $x^2 + 2y^2$ and have left

$$r = s^2 + 2t^2$$

Now r is not a prime greater than 1, since it would then be a 2-good prime factor and should have been removed. If $r = 1$, then $x^2 + 2y^2$ has no 2-bad prime factors. If r is not prime, then it has at least two prime factors, both of which are 2-bad since they have not been removed. This proves the lemma.

Lemma 1 also works for the new problem, that is, Lemma 1b.

Lemma 1b. If p is prime and has a multiple $x^2 + 2y^2$, where x and y have no common factor, then there is a number $a^2 + 2b^2$ which is less than p^2 and divisible by p.

The proof is by making a and b congruent to x and y (or $-x$, $-y$) just as in Lemma 1.

Having proved Lemmas 1b and 2b, we can apply the method of infinite descent and obtain a factor theorem for $x^2 + 2y^2$. If x and y have no common factor, then $x^2 + 2y^2$ has no 2-bad prime factors. In other words, every prime factor of $x^2 + 2y^2$ is of the form $a^2 + 2b^2$.

We need not tax our minds over this proof by infinite descent. It can be copied almost word for word from the proof for hypotenuses. All the hard work is behind us, once we have Lemmas 1b and 2b.

Try the factor theorem on a few examples. Let $x = 2$, $y = 3$. Then $x^2 + 2y^2 = 22 = 11 \times 2$. Sure enough, $11 = 3^2 + 2 \times 1^2$, and $2 = 0^2 + 2 \times 1^2$. Or let $x = 2$, $y = 7$. Then $x^2 + 2y^2 = 102 = 2 \times 3 \times 17$. Now $2 = 0^2 + 2 \times 1^2$, $3 = 1^2 + 2 \times 1^2$, and $17 = 3^2 + 2 \times 2^2$.

What about $x^2 + 3y^2$, $x^2 + 5y^2$, and so on? The multiplication theorem is true no matter what number is multiplied by y^2. Call this number n. Then if

$$g = a^2 + nb^2$$

and

$$h = c^2 + nd^2$$

we have

$$\begin{aligned} gh &= a^2c^2 + nb^2c^2 + na^2d^2 + n^2b^2d^2 \\ &= (ac + nbd)^2 + n(ad - bc)^2 \\ &= (ac - nbd)^2 + n(ad + bc)^2 \end{aligned}$$

Also, if p is prime and

$$p = a^2 + nb^2$$
$$pq = x^2 + ny^2$$

then it follows just as before that either $ax + nby$ or $ax - nby$ is divisible by p, and from this we can prove that $q = u^2 + nv^2$, and finally we arrive at Lemma 2c.

Lemma 2c. If x and y have no common factor, then $x^2 + ny^2$ cannot have just one n-bad prime factor. It has either none or at least two.

An n-bad prime is one which is not the sum of any square with n times another square.

Apparently we are going to prove a factor theorem for $x^2 + ny^2$, just like the ones for $x^2 + y^2$ and $x^2 + 2y^2$. But let us be wary. Suppose that $n = 5$, and $x = y = 1$. Then

$$x^2 + 5y^2 = 1^2 + 5 \times 1^2 = 6$$

Now, the factors of 6 and 2 and 3, neither of which is equal to $a^2 + 5b^2$ for any choice of a and b. Therefore the factor theorem is not true for $x^2 + 5y^2$. How is it, then, that we seem to be able to prove such a theorem? It there something wrong with the proof by infinite descent? Or did I slip something by in asserting Lemma 2c?

The flaw is not hidden in the inner recesses of the proof. It is lying right out in the open, like Dupin's stolen letter. It has to do with Lemma 1. To prove the factor theorem, we need Lemma 1c.

Lemma 1c. If p is prime and has a multiple $x^2 + ny^2$, where x and y have no common factor, then there is a number $a^2 + nb^2$ which is less than p^2 and divisible by p.

Now, the method of finding a and b is to make them congruent to x or $-x$ and y or $-y$. In this way they can both be made smaller than ½p, and $a^2 + nb^2$ is congruent to 0. But does it follow that $a^2 + nb^2$ is less than p^2?

We know that a^2 and b^2 are both less than $\frac{1}{4}p^2$. Therefore $a^2 + nb^2$ is less than $\left(\frac{n+1}{4}\right)p^2$. If n is 1 or 2, this is all we need. But if n is 5 or more, there is no guarantee that $a^2 + nb^2$ is less than p^2. Thus Lemma 1c is actually not true. This spoils the proof and explains why the factor theorem doesn't work if n is 5 or more.

If $n = 3$, we must be very careful. Then $\left(\frac{n+1}{4}\right)p^2$ is the same as p^2, so that apparently $a^2 + nb^2$ is less than p^2. This is true if p is an odd prime, for then a and b are both *less* than $\frac{1}{2}p$ (which is a fraction). But if $p = 2$, a and b may be *equal* to $\frac{1}{2}p$, which is 1. In fact

$$1^2 + 3 \times 1^2 = 4 = 2^2$$

Since the only multiple of 2 *smaller* than 2^2 is 2, which is *not* equal to any $a^2 + 3b^2$, we see that Lemma 1c is not true for $n = 3$ and $p = 2$. But it is true for $n = 3$ if p is not 2.

This partial result means that the factor theorem can still be proved, provided we can find a separate way of eliminating the factor 2. This can be done by a variation on Lemma 2c, in which we use the number 4 instead of the prime p.

Suppose that x and y have no common factor, and $x^2 + 3y^2$ is even. Then x and y are both odd. Either they are congruent, modulo 4, or they are not. Suppose they are. Then $x + 3y$, $x - y$ are both multiples of 4

$$x + 3y = 4u$$
$$x - y = 4v$$

It follows that $x = u + 3v$, $y = u - v$, and $x^2 + 3y^2 = 4(u^2 + 3v^2)$.

If x and y are not congruent, then we can put

$$x - 3y = 4u$$
$$x + y = 4v$$

and once again it turns out that $x^2 + 3y^2 = 4(u^2 + 3v^2)$.

We now start over with $u^2 + 3v^2$. But u and v have no common factor (x and y would have it too) and therefore either $u^2 + 3v^2$ is odd, or both u and v are odd. We can repeat the process until we have an odd number $r^2 + 3s^2$ which is just $x^2 + 3y^2$ divided by a power of 4.

The correct factor theorem for $n = 3$ is that if x and y have no common factor, then $x^2 + 3y^2$ has no *odd* 3-bad prime factor. To prove it, suppose that $h = x^2 + 3y^2$ and p is an odd prime 3-bad factor of h. First eliminate factors of 4, as shown above, until you get an odd number $r^2 + 3s^2$ which still has the factor p. Then apply Lemma 2c to find another 3-bad factor, q. Since $r^2 + 3s^2$ is odd, q is odd. Therefore Lemma 1c does work and there is a number $h' = x'^2 + 3y'^2$, which is less than h and divisible by either p or q. The infinite descent is off and running.

Euler's Missing Lemma

We are going to look at two theorems that have to do with the formulas $x^2 + 2y^2$ and $x^3 + y^3$. The question is whether either formula can be made a cube, by choosing integers x and y. Some authors say that Euler proved the two theorems in his *Elements of Algebra* published in 1770. But both proofs require a lemma that he takes for granted. In the first edition of this book, I carelessly stated that he had proved the lemma elsewhere, but apparently the proof is nowhere in his works. Still, other authors who call the proofs in the *Elements* incomplete don't bother to name the "other mathematicians" who filled in the missing steps, thus showing that they consider the missing lemma a straightforward consequence of things already known.

The lemma, which I shall call "Euler's Missing Lemma," is

stated and proved in Appendix II on p. 231. Here I shall continue with the two theorems, taking the lemma for granted.

Twenty-five and Twenty-seven

Euler found a peculiar use for the factor theorem on $x^2 + 2y^2$. It is an odd fact that the cube of 3 (27) and the square of 5 (25) lie very close together. He set out to find whether there are other solutions of

$$x^2 + 2 = u^3$$

besides $x = 5$, $u = 3$.

Now this equation can be written

$$x^2 + 2y^2 = u^3 \qquad \text{where } y = 1$$

The missing lemma then tells us that one can choose a and b so that (apart from a possible minus sign)

$$x = a^3 - 6ab^2$$
$$y = 3a^2b - 2b^3$$

The second equation means that b is a factor of y. But we want y to be 1. Therefore b must be 1, and we have

$$1 = 3a^2 - 2 \qquad \text{or } a = 1$$

Therefore the *only* solution to $x^2 + 2 = u^3$ has (from the first equation)

$$x = 1^3 - 6 \times 1 \times 1^2$$
$$= -5$$

which can just as well be made $x = 5$ since x^2 comes out the same. This is the original solution. There are no others.

The Sum of Two Cubes

A very similar method enabled Euler to prove a much more important theorem; a cube cannot be the sum of two cubes. This is part (but by no means all!) of Fermat's Last Theorem (see Chapter 5).

Suppose that

$$x^3 + y^3 = z^3$$

and that x, y, z have no common factor, and that z is even (which means x and y are odd). Euler showed by infinite descent that this is impossible.

Since x and y are odd, their sum and difference are even. Let p = half the sum, q = half the difference. Then

$$x = p + q$$
$$y = p - q$$

$$\begin{aligned} z^3 = x^3 + y^3 &= (p^3 + 3p^2q + 3pq^2 + q^3) + (p^3 - 3p^2q + 3pq^2 - q^3) \\ &= 2(p^3 + 3pq^2) \\ &= 2p(p^2 + 3q^2) \end{aligned}$$

I leave it to you to prove the following remarks.

Remark 1: p and q have no common factor.

Remark 2: $p^2 + 3q^2$ is odd.

Remark 3: q is not divisible by 3. (For this you must use the fact proven earlier in this chapter, that every cube is congruent to 0, 1, or -1, modulo 9. Hint: If q is divisible by 3, then $x \equiv y$, modulo 3, and hence $x^3 \equiv y^3$, modulo 9.)

Remark 4: p and $p^2 + 3q^2$ have no prime factor in common, except possibly 3.

We now apply the unique factorization theorem, just as we did in the theorems about squares (Chapter 5). If p is not divisible by 3, then $2p$ and $p^2 + 3q^2$ have no factor in common (Remark 4).

Since their product is a cube, they must both be cubes. Then we can apply the Missing Lemma (Remarks 1 and 2) and find that $q = 3m^2n - 3n^3$. But this contradicts Remark 3.

Therefore p must be divisible by 3. Say $p = 3w$. Then

$$z^3 = 18w(q^2 + 3w^2)$$

By Remarks 1, 2, and 3, we now see that $18w$ and $q^2 + 3w^2$ have no factors in common. Therefore each of them is a cube. Now the Missing Lemma tells us that $w = 3m^2n - 3n^3$, and therefore

$$\begin{aligned} 18w = 54(m^2n - n^3) &= 27 \times 2n \times (m^2 - n^2) \\ &= 27 \times 2n \times (m + n) \times (m - n) \end{aligned}$$

But $18w$ is a cube, and so is 27. Therefore $2n \times (m + n) \times (m - n)$ is a cube. I leave it to you to show that no two of the three numbers $2n$, $m + n$, $m - n$ have a factor in common. Hence all three are cubes. Let us write

$$2n = c^3, \qquad m + n = a^3, \qquad m - n = (-b)^3$$

Then

$$a^3 + b^3 = c^3$$

and c is even, and a, b, c have no factor in common. Therefore we can start over with a, b, c in place of x, y, z. This leads to an infinite sequence of triplets.

The Lemma does not say whether m is greater or smaller than n. (They cannot be equal, since then p and q would have a common factor.) Thus b may turn out either positive or negative. This causes no difficulty, when the sequence is extended, since there was nothing in the argument that required y to be positive.

If negative numbers are allowed, we must be careful about the "infinite descent." An infinite descent is impossible only if it consists of *positive* numbers. Therefore, consider the quantity

$a^6 + b^6 + c^6$. This is certainly positive, and I leave it to you to show that it is less than $x^6 + y^6 + z^6$. (Hint: Express both z^3 and $a^6 + b^6 + c^6$ in terms of m and n.) Therefore the numbers $x^6 + y^6 + z^6$, $a^6 + b^6 + c^6$, etc., form an infinite descending sequence of *positive* numbers, which is impossible.

I made two assumptions about the original x, y, and z. I supposed that they have no common factor. But if they had a common factor, it could be removed, and the equation $x^3 + y^3 = z^3$ would still be true. This could be continued until there was no common factor.

I also supposed that z is even. But this can always be arranged. Certainly one of the three numbers, x, y, z, is even, for the sum of two odd numbers is never odd. We can just call the even one z, and replace one of the others by its negative if necessary. Thus, the equation

$$22 + 7 = 29$$

can just as well be written

$$(-7) + 29 = 22$$

And if 7 were a cube, -7 would also be a cube. (This is true of cubes, but not of squares.)

Therefore *any* triplet satisfying

$$x^3 + y^3 = z^3$$

can be used to start an infinite descent. This proves that no such triplet exists.

8.

A new wind

Wilson's Theorem

In the eighteenth century some mathematicians discovered a curious test for prime numbers. Multiply together all the numbers less than n, and add 1. If the result (call it W_n) is divisible by n, then n is prime. Otherwise, n is not prime. Thus, $W_4 = 7$, $W_5 = 25$. So 5 is prime and 4 is not.

This is hardly a practical test, for large numbers n. If n is 97, it would take days to find W_n, which would be many digits long. The sieve of Eratosthenes is much easier.

However, Wilson's Theorem (as the test is called)* is a remarkable fact. It is usually stated with the help of an exclamation point, which has a special meaning in mathematics. The expression "4!" is read "factorial 4" and stands for the product of the first four integers.

$$4! = 1 \times 2 \times 3 \times 4 = 24$$

Just so, "$(n - 1)!$" means the product of the first $n - 1$ integers. So Wilson's Theorem says that

$$(n - 1)! \equiv -1 \qquad (\text{mod } n)$$

if n is prime, but not otherwise. (Even after years of practice, I can't read an equation involving factorials without getting a little bit excited.)

Half of Wilson's Theorem is easy to

* See note in Appendix 1.

prove. If n is not prime, it has a factor r, less than n and greater than 1. But r is a factor of $(n-1)!$, which is the product of all numbers less than n. Hence r cannot go into $(n-1)!+1$, and therefore n cannot.

Proving the other half, which says that if n is prime then it does go into $(n-1)!+1$, was too hard for anyone in the seventeenth century, even Fermat. There is a very simple proof, discovered by Euler in 1773, but it contains a subtle new idea.

Let us call two numbers *inverse* if their product is congruent to 1, modulo n. Then it can be shown (see below) that each number from 1 to $n-1$ has just one inverse below n, and that the only numbers below n which are their own inverses are 1 and $n-1$. Once you grant that, you must see that all the numbers from 2 to $n-2$ can be paired up, each with its own inverse. In order to multiply them together, we can first multiply together each pair of inverses, getting a number congruent to 1, modulo n. When all these products are multiplied, the answer will still be congruent to 1, modulo n, because the product of any number of 1's is just 1. Then, to find $(n-1)!$, we must multiply by 1 and by $n-1$, which were not included in the pairs. Clearly, the final answer is congruent to $n-1$, or to -1. Hence $(n-1)!+1$ is divisible by n. For example, if $n=7$, we pair 2 with 4, and 3 with 5, thus

$$\begin{aligned}(n-1)! &= 1 \times 2 \times 3 \times 4 \times 5 \times 6 \\ &= (2 \times 4) \times (3 \times 5) \times 1 \times 6 \\ &\equiv 1 \times 1 \times 1 \times (-1) \\ &\equiv -1 \qquad \pmod 7\end{aligned}$$

The existence of inverse pairs follows from the assumption that n is prime. If a is less than n, then a and n have no common factor greater than 1. Hence (GCF theorem, Chapter 3) there are two numbers b and m, such that

$$ab + mn = 1$$

but then

$$ab \equiv 1 \qquad (\text{mod } n)$$

Now b may be greater than n or negative, but it is certainly congruent to some positive number less than n, and so it may be replaced by that number, which is evidently inverse to a, according to the definition given above. So each a has an inverse.

But a cannot have *more* than one inverse between 0 and n. For if b and c are both inverse to a, then

$$ab \equiv ac \qquad (\text{mod } n)$$

and hence

$$b \equiv c \qquad (\text{mod } n)$$

since division is permitted when the modulus is prime. Therefore b and c are equal or else they are not both between 0 and n.

Finally, if a is its *own* inverse, then $a^2 - 1$ is divisible by n. That is, $(a - 1)(a + 1)$ is divisible by n, which is impossible if a is any number from 2 to $n - 2$. This completes the proof of Wilson's Theorem.

The Language of Residues

In this proof it is a nuisance to worry about whether a number is between 0 and n. We can eliminate that trouble by speaking not of numbers but of new creatures called *residues*. Having chosen n as the modulus, we say that there are just n residues, which we call 0, 1, 2, up to $n - 1$. But each residue has many names. Thus 3, $n + 3$, and $2n + 3$ are all names of the same residue. Instead of reading

$$2 \equiv 7 \qquad (\text{mod } 5)$$

as "the number 2 is congruent to the number 7, with 5 as modulus," we read it as "2 and 7 are names of the same residue, with 5 as modulus." And instead of reading

$$2 + 4 \equiv 1 \pmod{5}$$

as "the sum of the numbers 2 and 4 is congruent to the number 1," we read it as "the sum of the residues 2 and 4 *is* the residue 1."

The laws governing congruences (see Chapter 6) make it possible to simplify our language in this way without ever falling into a contradiction on account of having many names for the same residue. In fact, the new language will seem hardly different from the old. The advantage is that whereas there are infinitely many numbers, there are just n residues, and this fact is the basis of many proofs.

In the new language we should say that the product of any residue with its inverse *is* 1, and therefore, when all $n - 1$ residues (excluding 0) are multiplied together, everything cancels out except the two residues, 1 and -1, which are inverse to themselves. Thus the final product *is* the residue -1. Details must be filled in as before, but the qualification "between 0 and n" can be dropped.

Let us use this new language in a proof of Fermat's Theorem, also due to Euler. Let n be prime and x be a residue, not 0. Write the residues 1, x, x^2, and so on. Since there are only n residues, the sequence must eventually repeat.

Now the first residue that is repeated must be the first residue in the sequence, that is, 1. This can best be seen by examining the way it is not true, if n is *not* prime. Let $n = 10$, $x = 2$. The sequence of powers is

$$1, 2, 4, 8, 6, 2, 4, 8, 6, 2, \ldots$$

Here 2 is the first residue repeated. Clearly, this is possible only because there are two ways of arriving at 2: by doubling 1, or by doubling 6. But that is just what cannot happen if n is prime; then there is only one residue which can be multiplied by x to give a desired result. Hence no power of x can be repeated, if n is prime, until the power before it has been repeated. So 1 must be repeated first of all.

Let x^r be the lowest power of x that comes out to be 1. Then the sequence repeats itself after each r members. For example, if $n = 7$, $x = 4$, the sequence goes 1, 4, 2, 1, 4, 2, etc. Then $r = 3$.

Now the r residues $1, x, \ldots, x^{r-1}$ form an unending chain which we may call an x-chain because each residue is x times the previous one. But we can also start from any nonzero residue, a, and generate an x-chain $a, ax, ax^2, \ldots, ax^{r-1}$. This will also have r members and will connect at the ends, since $ax^r \equiv a$. (Note that we multiply by x, not by a.) In this way every residue except 0 belongs to some x-chain.

Some of the x-chains may overlap. But it is easily seen that if two chains overlap at all, they must overlap completely, so that they contain all the same residues, only with different starting points. First suppose that we start a chain at a, and that one of the "links" of the chain is b. Then if we start a new chain at b, it is perfectly obvious that we get all the links of the old chain and no others; both chains are continued from b in the same manner, by multiplying by x. The part of the old chain that came before b will appear in the new chain after a, which must come sooner or later.

Now consider any two x-chains, one starting at a and one at c. Perhaps they are completely separate. But if they meet anywhere, say at b, then we can start a third chain at b. The third chain is identical to each of the others, as shown above. Therefore the chains starting at a and c are identical.

It follows that we may find separate chains, each containing r residues, which include among their links all the $n - 1$ nonzero residues. Hence $n - 1$ is a multiple of r. But we know that $x^r \equiv 1$; therefore $x^{n-1} \equiv 1$, and Fermat's Theorem is proved.

This is a new twist. We have proved a relation between

$n - 1$ and r, not by manipulating the *numbers* $n - 1$ and r, but by manipulating certain *objects* (the residues modulo n) and counting these objects as they fall into certain categories. The language of residues is very handy because it permits us to manipulate and count the same objects.

Euler, although he originated this type of proof, used the old language in all his papers. This made the proof more difficult because he had to keep in mind two sets of numbers: one (the numbers less than n) that could be counted, and another (the powers of x, related to the first set by congruence) that could be manipulated, as in generating chains. The idea of melting down into one residue all the numbers congruent to one another, and performing arithmetic on the residues rather than on the numbers, appears first in Carl Friedrich Gauss's book *Disquisitiones Arithmeticae,* which he finished in 1801, at the age of twenty-four. (The title means "Arithmetical Researches," but I shall call it just *Disquisitions.*)

A Masterpiece

Disquisitions is remarkable in several ways. First, it is one of only two book-length works ever written by the man recognized as the greatest mathematician of the last two centuries and perhaps of all time. Second, it combines the thoroughness of a textbook with the excitement of original discovery. This is most unusual. A scientific discovery is usually announced in a paper, that is, a short article published in a professional journal. Each related discovery goes into a new paper. When enough discoveries have been made in a particular subject—perhaps after fifteen or twenty years of investigation by many scholars—some one will collect the new information, organize it into a science, eliminate the unnecessary parts, put the important parts in their proper relation,

Carl Friedrich Gauss, 1777–1855

and explain it in a book. This is what Euclid did for geometry. Until such a book has appeared, it is extremely difficult for anyone outside the subject to learn about the new discoveries. Unless one is oneself one of the leaders in this research, one will not know which paper to read first or where any one paper fits into the whole field. Most of Euler's discoveries in number theory were not brought together in any book for decades after they were published.

As time goes on, more books are written, in order to explain the earlier books to those who could not understand them. The later books are easier to understand, but less and less exciting, just as a railroad trip through the Far West is easier but less exciting than a journey by covered wagon during the wars with the Plains Indians. Later books also tend to give wrong impressions of how a discovery was made.

The most exciting and often the clearest exposition of a subject is apt to be in a book written by one of the original discoverers, if he happens to be good at writing books. Such a book is usually a summary of the author's own papers from years back, and of the papers of others, with some details changed or omitted so that the whole is done from a single point of view. Gauss's *Disquisitions* is exceptional in that most of it is material newly discovered by him, and never before published in any form. And yet all this new, raw information is woven into a work of several hundred pages, as thorough and well-planned as any textbook. Thus it combines the excitement and brilliance of Euler's scattered mathematical papers with the logical perfection of Euclid's *Elements*. There must be very few examples in history of so finished a work, on such a large scale, expounding discoveries so new in the author's own mind.

In the third place, *Disquisitions* was one of the first modern books devoted mainly to the theory of numbers. (I cannot say it was *the* first, since Adrien Legendre's *Theory of Numbers* appeared at about the same time.) During the eighteenth century, mathematicians spent most of their time on algebra and calculus, which were developed much further than number theory. Even for Euler, the problems of number theory were an occasional distraction which intervened from time to time between his other researches. Legendre and Gauss were the first to see number theory as a unified subject worthy of systematic presentation.

A Little Algebra

An example of the subordination of number theory to algebra may be seen in Euler's first proof of Fermat's Theorem. The proof I have described above was actually the second, dating from the 1750's. Euler had already proved the theorem in 1736 by a quite different method, depending on the *binomial theorem* of algebra.

You may have learned, or you can easily see for yourself, that the square of $x + y$ is $x^2 + 2xy + y^2$. And the cube of $x + y$ is $x^3 + 3x^2y + 3xy^2 + y^3$. As we take higher powers, we get higher numbers appearing in the expression; thus

$$(x + y)^6 = x^6 + 6x^5y + 15x^4y^2 + 20x^3y^3 + 15x^2y^4 + 6xy^5 + y^6$$

The numbers 1, 6, 15, 20, 15, 6, 1 are called *binomial coefficients* of degree 6. The binomial coefficients of any degree can be found simply by multiplying $x + y$ by itself the right number of times. But there is also a formula which gives them directly. If the degree is n, then the binomial coefficients are 1, n, $\frac{n(n-1)}{2!}$, $\frac{n(n-1)(n-2)}{3!}$, and so on until you get to $\frac{n(n-1)(n-2)\ldots\times 2\times 1}{n!}$, which is equal to 1. It looks as if some of these coefficients might be fractions, but in fact they always come out whole numbers. Try working them out for $n = 6$.

This rule for finding the coefficients is called the binomial theorem, and it was well known in Euler's day. To prove Fermat's Theorem from it, Euler pointed out that if n is prime, then the factor n in the numerator cannot be cancelled by any of the smaller factors in the denominator. Therefore all but the first and last coefficients will be divisible by n. Thus, the binomial coefficients of degree 5 are 1, 5, 10, 10, 5, 1, and those of degree 7 are 1, 7, 21, 35, 35, 21, 7, 1. Consequently, if x and y are integers, then

$$(x + y)^n \equiv x^n + y^n \pmod{n}$$

provided that n is prime. The missing terms on the right are all congruent to zero. (See how this fails for $n = 6$.)

Now suppose that x and y both obey Fermat's Theorem, with n as exponent. That is,

$$x^n \equiv x \pmod{n}$$

and

$$y^n \equiv y \pmod{n}$$

It then follows from Euler's congruence that

$$\begin{aligned}(x + y)^n &\equiv x^n + y^n \\ &\equiv x + y \pmod{n}\end{aligned}$$

That is, $x + y$ also obeys Fermat's Theorem. But now Fermat's Theorem can easily be proved. It is obviously true if $x = 1$, since $1^n = 1$. Therefore it is true for $1 + 1$, or 2. And therefore it is true for $2 + 1$, or 3. Thus it is true for all numbers.

Note that this proof uses induction on x but not on n. The logical chain runs from $x = 1$ to $x = 2$ and so on, but there is no chain running from one value of n to another. The proof applies to each prime value of n, independently of other values.

Note also that this proof, unlike the later one given above, does not use the counting of residues. But it leans heavily on a prior knowledge of algebra. In fact, the crucial part is the proof (which I have left out) of the binomial theorem itself.

A Lot More Algebra

Even more striking is the contrast between Euler's proof of Wilson's Theorem, which began this chapter, and an earlier proof, which was produced not by Euler but by Joseph Louis Lagrange, in 1771.

Lagrange was born in 1736, when Euler was twenty-nine. Between them they were the leading mathematicians of the eighteenth century (Gauss belonged to the nineteenth century) and the range of subjects they studied is almost identical. Some of the most important methods in mathematical physics depend on "Euler-Lagrange" equations. Typically, Euler would discover the germ of an idea in the course of solving a particular problem. He would then go on to other matters and Lagrange would take up the idea, generalize it, show how it could be used for a large class of problems, and exhibit its relation to other ideas. Lagrange was not just an imitator who followed the paths Euler had laid out. He added so much that he must be rated as Euler's equal, but he is most distinguished by his desire to unify and generalize where Euler was content with the fragmentary and particular. His favorite subject was algebra, in which many statements about individual numbers can be united into one equation with letter variables.

Lagrange approached Wilson's Theorem by unifying it with Fermat's Theorem. He considered the product of n consecutive numbers, beginning with x. Let us call this product $L_n(x)$. Thus

$$L_n(x) = x(x + 1)(x + 2) \dots (x + n - 1)$$

If we let n be a prime number and do the multiplication algebrai-

Joseph Louis Lagrange, 1736–1813

cally, we find something surprising. We have

$$\begin{aligned}
L_3(x) &= x(x+1)(x+2) \\
&= x^3 + 3x^2 + 2x \\
&= x^3 - x + 3(x^2 + x) \\
L_5(x) &= x(x+1)(x+2)(x+3)(x+4) \\
&= x^5 + 10x^4 + 35x^3 + 50x^2 + 24x \\
&= x^5 - x + 5(2x^4 + 7x^3 + 10x^2 + 5x) \\
L_7(x) &= x(x+1)(x+2)(x+3)(x+4)(x+5)(x+6) \\
&= x^7 + 21x^6 + 175x^5 + 735x^4 + 1624x^3 + 1764x^2 + 720x \\
&= x^7 - x + 7(3x^6 + 25x^5 + 105x^4 + 232x^3 + 252x^2 + 103x)
\end{aligned}$$

In each case the difference between $L_n(x)$ and $x^n - x$ is some expression multiplied by n. That is, $L_n(x)$ is *identically* congruent to $x^n - x$, modulo n. The word *identically* means that it is not necessary to substitute a value for x in order to show that $L_n(x) - (x^n - x)$ is divisible by n. The factor n can be removed directly from the expression in x.

Note that "identically" means *more* than "always". Fermat's Theorem, for example, says that x^3 is *always* congruent to x, modulo 3, that is, for *any* numerical value of x, the value of $x^3 - x$ will be a number divisible by 3. But it is still necessary to give x a value; x^3 is not *identically* congruent to x, because the expression $x^3 - x$ cannot be divided by 3 just as it stands. On the other hand, $L_3(x) = x^3 + 3x^2 + 2x$ is *identically* congruent to $x^3 - x$, because the difference $3x^2 + 3x$ can be divided by 3 to obtain $x^2 + x$, without giving x a numerical value at all.

Now Lagrange fastened on the idea that Fermat's and Wilson's theorems both follow from the fact that $L_n(x)$ and $x^n - x$ are identically congruent, for prime n. For although it is quite complicated to multiply out the expression $L_n(x)$, it is clear that the last term of the answer will always be $(n-1)!x$. (Prove this by induction on n; remember that $L_n(x) = (x + n - 1) \times L_{n-1}(x)$.) Therefore the difference $L_n(x) - (x^n - x)$ will contain a term $(n - 1)!x + x$ or $[(n - 1)! + 1]x$. (Thus $L_5(x) - (x^5 - x) = 10x^4 + 35x^3 + 50x^2 + 25x = 10x^4 + 35x^3 + 50x^2 + (4! + 1)x$.) Now if $L_n(x)$ and $(x^5 - x)$ are *identically* congruent, then each power of x that appears in the difference must be multiplied by a number divisible by n. Therefore $(n - 1)! + 1$ must be divisible by n, and Wilson's Theorem is proved.

On the other hand, if two expressions are identically congruent, then their values will certainly be congruent whenever x is given a particular value. Now $L_n(x)$ is *always* congruent (not identically!) to 0, modulo n. For if x is any integer, *one* of the n numbers from x to $x + n - 1$ must be divisible by n. Hence the product of the n numbers is divisible by n. So, if $L_n(x)$ is identically (and therefore always) congruent to $x^n - x$, it follows that $x^n - x$ is always congruent to 0. This proves Fermat's Theorem.

Lagrange's Theorem

It remains to show that $L_n(x)$ and $x^n - x$ are indeed identically congruent provided that n is prime. In Lagrange's original memoir, he proved this directly and then deduced the two theorems. But it is simpler (and really no different from his procedure) to assume that Fermat's Theorem has been proven by Euler's method and thence to show the identical congruence, of which Wilson's Theorem is a consequence.

By the argument of our next to the last paragraph, we see that $L_n(x)$ is *always* congruent to 0, and therefore to $x^n - x$ if Fermat's Theorem is granted. But this does not prove that $L_n(x)$ and $x^n - x$ are *identically* congruent. For that we must draw on the theory of polynomials.

A polynomial is an expression like $x + 5$ or $5x^2 - 2x - 7$ or $3x^3 - 6x$; that is, a sum of different powers of x, each multiplied by some number. If x is given a value, the polynomial takes on a value. For some values of x, the polynomial may take the value 0. These values of x are called *roots* of the polynomial. Thus, -5 is a root of $x + 5$; $7/5$ and -1 are roots of $5x^2 - 2x - 7$; 0 is a root of $3x^3 - 6x$.

A basic theorem of algebra states that unless a polynomial is *identically* zero, it cannot have more roots than the highest power of x which appears in it. This power is called its *degree.* A polynomial is not identically zero unless every power of x cancels out separately. Thus $4x - 3 + x - 5x + 3$ is identically zero. But $x + 5$ is not identically zero, and its degree is 1; therefore it has no more than the one root, -5. The degree of $3x^3 - 6x$ is 3; therefore it may have two roots besides 0, but no more than that.

This theorem (which you will probably learn in a good high-school course in algebra) was known since the seventeenth century. One proof depends on showing that if r is a root, then the polynomial can be divided by $x - r$ without remainder. So, if there were more than s roots, the polynomial would have more than s factors $x - r_1$, $x - r_2$, etc. Therefore it would contain higher powers of x than the sth.

Lagrange saw that this proof could be carried over from equations to congruences, and in 1768 he came up with this theorem: If a polynomial of degree s is not identically congruent to 0 (with a prime modulus) then there cannot be more than s different residues x for which the value of the polynomial is congruent to 0. (Note that two numbers that are congruent are counted only once as a residue.) This theorem will play so important a part in the rest of the book that I shall refer to it as Lagrange's Theorem. It is exactly the same as the high-school algebra theorem except that residues are used in place of numbers, and it is proved in the same way. Let us skip the proof, since this book is not supposed to teach you algebra!

Lagrange's Theorem can help us decide whether two polynomials may be always congruent, without being identically congruent. Suppose that a polynomial is always, but not identically, congruent to zero, modulo n. Then it has n "different" roots, the residues from 1 to n. This is consistent with Lagrange's Theorem only if the degree of the polynomial is at least n. We may draw the following conclusion: if a polynomial *of degree less than n* (where n is prime) is always congruent to zero, modulo n, then it must be identically congruent to zero, modulo n.

This conclusion is not contradicted by $x^n - x$, because its degree is n. Therefore it is allowed to be always congruent, but not identically congruent, to zero. The same is true of $L_n(x)$, which also has degree n. But the difference $L_n(x) - (x^n - x)$ is another story. The first term in $L_n(x)$ is obviously x^n. This cancels the first term of $x^n - x$ so that the degree of $L_n(x) - (x^n - x)$ is only $n - 1$. (Look at the examples given earlier.) And yet this polynomial is always congruent to zero, modulo n; hence it is identically divisible by n, and $L_n(x)$ and $x^n - x$ are identically congruent. It is crucial to note the role played in this proof by the *degrees* of the polynomials.

The Trumpet of a Prophecy

Now that we have established the identical congruence of $L_n(x)$ and $x^n - x$, Lagrange's proof of Wilson's Theorem is finally complete. The remaining steps were given a few pages back and we need not repeat them. If you have followed this proof, you are probably astonished that such a difficult, roundabout argument was discovered *earlier* than Euler's simple proof based on the pairing of inverses.* The key to this historical paradox is the advanced state of algebra in the eighteenth century. Most of the difficulty in Lagrange's proof resides in concepts and reasoning

* Euler actually gave a slightly less simple version which used the same principle, but depended on the existence of primitive roots (see Chapter 9). The version I gave on pages 154–155 is due to Gauss.

that belong entirely to algebra and were already well understood in his time. And remember that Lagrange himself was a pioneer in the most difficult algebraic problems. Once the algebra is under your belt, you can follow his proof of Wilson's Theorem in a few short steps.

Euler's proof, simple though it seems, was actually more difficult at the time because the technique of counting residues was entirely unfamiliar. Not only unfamiliar, it was an early breath of a new wind that began to blow over all mathematics only in the time of Gauss. For one thing, it involved the idea (although Euler himself didn't spell it out) that all the numbers congruent to each other according to the given modulus should be treated as a single entity, a sort of fictional number, which we have called a residue. In the nineteenth century many advances were made by introducing new entities into mathematics, and nowadays the procedure is commonplace. But in the eighteenth century it was unheard of. Mathematics was the science of quantity, and each number represented a quantity. You couldn't invent a new number or a new quantity; you could only find out new facts about the old ones.

Since the nineteenth century, mathematics has been the science of form and relation. That is, any abstraction, whether number or residue or what else, which is related to other abstractions according to definite patterns, is an object for mathematical study. One is continually introducing new abstractions, to study the form of the patterns they make. They need not represent quantities.

Another prophetic feature of "residue counting" is its emphasis on the *set* of all residues. In the past, mathematics had almost always dealt with one number at a time. This is especially true of algebra. Even though a statement about the letter x is meant to apply to all numbers, it applies to them *one at a time.* The statement can be understood by letting $x = 1$, or $x = 36$, and so on, but not by letting x be many numbers at once. And a statement of identity, such as that $x + 3x$ is identical to $4x$, does not really involve any numerical values of x, but only the single *expressions*, $x + 3x$ and $4x$.

In calculus, which began in the seventeenth century, things are a little different. There, the derivative of x^2 at $x = 10$ depends not only on the square of 10 but also on the squares of a great many numbers close to 10. However, the eighteenth-century mathematicians, partly for this very reason, were quite confused about the foundations of calculus, although they became adept in its methods.

If you look again at Euler's "chain" proof of Fermat's Theorem (not the one that uses the binomial theorem) you will see that the fact that $n - 1$ is divisible by r depends on the statement that the $n - 1$ nonzero residues fall into chains of r residues apiece. This statement involves all the residues in their relation to one another—all the residues at once, not each one in turn. Such an idea, simple as it is, was entirely foreign to the usual methods of eighteenth-century mathematics.

In the work of Gauss and those after him, this idea caught on and spread far beyond number theory. By the 1830's, the notion of a whole set of objects was being used to crack some of the most difficult problems of algebra, which had stumped even Lagrange, and to throw a new light on the basic ideas of calculus and infinite series. Today, set theory is regarded as the very foundation of mathematical ideas.

9.

Roots go deep

Euler's Toils

We are not done with Lagrange's Theorem. It will prove a mine of useful information about the residues of a prime modulus. Euler, for example, would have found it most useful in 1747. At that time he had already discovered that every factor of a primitive hypotenuse is also a primitive hypotenuse. (See Chapter 7.) In order to prove the famous statement of Fermat, that every upper prime is the sum of two squares, he had now only to show that an upper prime always has some multiple which is a primitive hypotenuse.

It occurred to Euler that if

$$p = 4n + 1$$

and is prime, it ought to be possible to find a number x such that $x^{2n} + 1$ is divisible by p. This would complete the proof, since x^{2n} and 1 are squares with no common factor besides 1. The reason Euler thought of using $x^{2n} + 1$ is that he knew, by Fermat's Theorem, that $x^{4n} - 1$ is *always* divisible by p, if x is an integer not divisible by p. Since $x^{4n} - 1$ is the product of $x^{2n} + 1$ and $x^{2n} - 1$, it is necessary only to show that there is some number x, not divisible by p, such that $x^{2n} - 1$ is *not* divisible by p. Then $x^{2n} + 1$ *will* be divisible.

It took Euler two years before he discovered a way to carry out this proof.

But it would have taken him no time at all if he had known Lagrange's Theorem. For there are $4n$ different residues modulo p besides zero. By Lagrange's Theorem not more than $2n$ of these can satisfy $x^{2n} - 1 \equiv 0$, and so the other $2n$ must satisfy $x^{2n} + 1 \equiv 0$, modulo p.

There is another way to solve this problem, which amounts to showing that if p is an upper prime, then -1 is congruent modulo p to some square. Since p is an odd prime, we can put $p = 2a + 1$. By Wilson's Theorem, $(2a)! \equiv -1$, modulo p. But since $2a \equiv -1$, $2a - 1 \equiv -2$, $2a - 2 \equiv -3$, etc., we can replace all the factors from $a + 1$ to $2a$ by negative factors from $-a$ to -1. Thus

$$\begin{aligned}(2a)! &\equiv a!(-a)\dots(-3)(-2)(-1) \qquad \pmod p\\ &= (-1)^a a! a!\\ &= (-1)^a a!^2\end{aligned}$$

If p is an *upper* prime, a must be even. Then $(-1)^a = 1$, and we have

$$a!^2 \equiv (2a)! \equiv -1 \qquad \pmod p$$

Thus $a!^2 + 1$ is divisible by p. For example, $2!^2 + 1$ is divisible by 5; $6!^2 + 1$ is divisible by 13. But $3!^2 + 1$ is not divisible by 7, for 3 is odd (and hence 7 is a lower number).

This trick was pointed out by Lagrange in 1771, when he published his proof of Wilson's Theorem.

Tiny Tables

Lagrange's Theorem also sheds light on the multiplication table for residues with a prime modulus. To show what I mean by "multiplication table," first consider the modulus 8, which is not prime. Since the product of any two odd numbers is odd, we can make a table for the odd residues alone.

The table says that $1 \times 3 \equiv 3$, $3 \times 7 \equiv 5$, $5 \times 5 \equiv 1$, and so on. The number 8 in the upper left-hand corner is the modulus.

8	1	3	5	7
1	1	3	5	7
3	3	1	7	5
5	5	7	1	3
7	7	5	3	1

A striking feature of this table is the string of 1's along the diagonal. This string reflects the fact that the product of any of the four residues by itself is congruent to 1. In other words, the square of any odd number is congruent to 1, modulo 8. This was already pointed out in Chapter 6. It seems to contradict Lagrange's Theorem, since there are now four different residues that satisfy $x^2 - 1 \equiv 0$. But remember that Lagrange's Theorem applies only when the modulus is prime.

If we make a table for the modulus 5, it will be the same size as the one for 8, since there are four residues if we leave out zero. But we know that it will be quite different in structure, because 5 is prime and therefore, according to Lagrange's Theorem, only two of the residues can satisfy $x^2 \equiv 1$. Suppose that r is one of the other two, so that r^2 is not congruent to 1, modulo 5. Then, of course, r is not 1. Also r^3 is not congruent to 1 because, if it were, r^4 would be congruent to r. But this is impossible because $r^4 \equiv 1$ by Fermat's Theorem.

It follows that the four residues $1, r, r^2, r^3$ are all different (for example, if $r \equiv r^3$, we could divide by r and get $1 \equiv r^2$, which is false). Since there are only four residues besides zero, we have shown that *every* number not divisible by 5 is congruent to some power of r. For this reason r is called a *primitive root* of 5. In fact, we have shown that 5 has two primitive roots, since r could have been either of the two residues that do not satisfy $x^2 \equiv 1$.

Let us verify this by making the table. As expected, the diagonal (from upper left to lower right) contains only two 1's, whereas in the modulus 8 table it contained four. The primitive roots are 2 and 3. Thus, if we start with 1 and take successive powers of 2, we get 1, 2, $2^2 = 4$, and $2^3 = 8 \equiv 3$. Or by taking powers of 3, we have 1, 3, $3^2 = 9 \equiv 4$, $3^3 = 27 \equiv 2$. Either way we get all four residues. But 4 is not a primitive root because its powers do not run through all four residues. Thus, after 1, 4, we have $4^2 = 16 \equiv 1$, $4^3 = 64 \equiv 4$, and so on.

5	1	2	3	4
1	1	2	3	4
2	2	4	1	3
3	3	1	4	2
4	4	3	2	1

Now consider the modulus 7. There are six residues besides zero. Of these, only two can satisfy $x^2 \equiv 1$, and only three can satisfy $x^3 \equiv 1$, by Lagrange's Theorem. Moreover, these two sets overlap because both congruences are satisfied if $x = 1$. Hence there can be no more than four residues for which either $x^2 \equiv 1$ or $x^3 \equiv 1$. (There is 1, and at most one other for which

7	1	2	3	4	5	6
1	1	2	3	4	5	6
2	2	4	6	1	3	5
3	3	6	2	5	1	4
4	4	1	5	2	6	3
5	5	3	1	6	4	2
6	6	5	4	3	2	1

$x^2 \equiv 1$, and at most two others for which $x^3 \equiv 1$.) Let r be one of the remaining two residues. Then r, r^2, r^3 are all different from 1. Therefore r^4 is different from 1 (if $r^4 \equiv 1$ then $r^6 \equiv r^2$ which contradicts Fermat's Theorem) and r^5 is different from 1 (since if $r^5 \equiv 1$ then $r^6 \equiv r$). Hence r is a primitive root of 7, and since r could have been either of two residues, there must be two primitive roots.

The multiplication table for the modulus 7 shows that 3 and 5 are primitive roots. The powers of 3 are 1, 3, $3^2 \equiv 2$, $3^3 \equiv 6$, $3^4 \equiv 4$, $3^5 \equiv 5$. But 2 is not a primitive root since its powers are 1, 2, $2^2 \equiv 4$, $2^3 \equiv 1$, $2^4 \equiv 2$, $2^5 \equiv 4$, and never 3, 5, or 6.

The existence of a primitive root makes it possible to simplify the multiplication table by changing the order of rows and columns. Instead of writing the residues in the order 1, 2, 3, 4, 5, 6, let us write them in the order in which they arise as powers of 3: 1, 3, 2, 6, 4, 5. Here every residue is three times the one before it. The multiplication table now looks much simpler.

This table contains the same information as the original one. For example, both tables say that $3 \times 4 \equiv 5$, modulo 7. But in this form each row is the same as the one above it, only shifted to the left. Thus you have only to know the top row in order to have the whole table at your fingertips.

7	1	3	2	6	4	5
1	1	3	2	6	4	5
3	3	2	6	4	5	1
2	2	6	4	5	1	3
6	6	4	5	1	3	2
4	4	5	1	3	2	6
5	5	1	3	2	6	4

The modulus 5 table can also be put in this form, by letting the top row read 1, 2, 4, 3. But the modulus 8 table cannot, since it has no primitive root.

Pears and Baskets

It would be nice to be able to say that every prime modulus has a primitive root. At first sight this seems to follow from Lagrange's Theorem, as we have seen when the modulus was 5 or 7. But the proof is harder when p (the modulus) is 1 more than a number with many factors. Let $p = 13$. Then there may be six residues that satisfy $x^6 \equiv 1$; four that satisfy $x^4 \equiv 1$; two, $x^2 \equiv 1$. This would use up all twelve nonzero residues. Does 13, then, have no primitive root?

If you look more carefully, you will see that some residues were counted twice. Those that satisfy $x^2 \equiv 1$ or $x^3 \equiv 1$ must satisfy $x^6 \equiv 1$ as well. Hence the six solutions of $x^6 \equiv 1$ must include all those of the other two equations. Then there are four solutions of $x^4 \equiv 1$, but two satisfy $x^2 \equiv 1$ and have already been counted. Six and two make eight, but altogether there are twelve nonzero residues, of which the remaining four must be primitive roots.

Evidently, if we are to prove that every prime modulus has a primitive root, we must find a systematic way of counting the residues that aren't primitive roots, without counting any twice. Euler studied this question in 1772 and finally published a "proof" which Gauss criticized in *Disquisitions* as being incomplete. Gauss gave a very beautiful proof, which I shall describe with minor changes.

Let p be a prime modulus. Then each residue besides zero has an *order*, which is the smallest power to which it need be raised to be congruent to 1. (Thus, if $p = 7$, the order of 2 is 3 since $2^3 \equiv 1$, but the order of 3 is 6 since 3^6 is the lowest power of 3 which is congruent to 1.) We may ask how many residues there are of each order.

The order of each residue is a factor of $p - 1$. This follows from the "chain" proof of Fermat's Theorem in the preceding chapter. Suppose, then, that n is some factor of $p - 1$. How many residues have order n? By Lagrange's Theorem, there cannot be more than n such residues, since they all must satisfy $x^n \equiv 1$.

But we can show that there are even fewer than n. Suppose r is one residue of order n. Then the powers of r all satisfy $x^n \equiv 1$. There are n distinct powers; if any two were congruent, the order would be less than n. Therefore (Lagrange's Theorem) the powers of r are the *only* solutions of $x^n \equiv 1$. But not all of them have order n. For if m and n have a factor in common, then the order of r^m is less than n. For example, if r has order 10, then r^6 obviously has order 5, since $(r^6)^5 = r^{30} = r^{10} \times r^{10} \times r^{10} \equiv 1$.

It follows that there cannot be more than $\phi(n)$ residues of order n. By $\phi(n)$ I mean the number of numbers less than n, having no factor in common with n. A handy way to express this is that $\phi(n)$ is the number of proper fractions in lowest terms, with denominator n. Thus, $\phi(10) = 4$ since there are four reduced proper fractions—$1/10$, $3/10$, $7/10$, $9/10$—with denominator 10. If r has order 10, then r, r^3, r^7, r^9 are the *only* residues of order 10. For example, r^6 has order 5 as shown above, and this is clearly connected with the fact that $6/10$ is not in lowest terms.

We have not proved that there *are* $\phi(n)$ residues of order n. There may not be any. But there cannot be more than $\phi(n)$. It is as though the residues had been dropped into different orders like pears into baskets. We do not know that every basket has received pears. We only know that each basket can hold no more than a certain number of pears.

But now we shall show that all the baskets put together can hold only as many pears as we have handed out. This will surely prove that all the baskets have been filled. In other words, we shall show that by adding together $\phi(n)$ (the sizes of the "baskets") for all the numbers n which are factors of $p - 1$, we get exactly the sum $p - 1$, which is the number of residues besides zero (the number of "pears").

The sum of the $\phi(n)$'s is just the number of proper reduced fractions with denominator a factor of $p - 1$. But any such fraction can be expressed with denominator $p - 1$. And any fraction with denominator $p - 1$ can be reduced to lowest terms in only one way. Thus the sum of the $\phi(n)$'s is just the number of proper fractions (not reduced) with denominator $p - 1$. But there are only $p - 1$ such fractions.

For example, let $p = 13$, $p - 1 = 12$. The 12 proper fractions are $1/12$, $2/12$, $3/12$, $4/12$, $5/12$, $6/12$, $7/12$, $8/12$, $9/12$, $10/12$, $11/12$, $12/12$. Reduced to lowest terms (and rearranged) they are 1, $1/2$, $1/3$, $2/3$, $1/4$, $3/4$, $1/6$, $5/6$, $1/12$, $5/12$, $7/12$, $11/12$. These 12 fractions include all reduced proper fractions with denominator 1, 2, 3, 4, 6, or 12. This expresses the fact that

$$\begin{aligned} 12 &= 1 + 1 + 2 + 2 + 2 + 4 \\ &= \phi(1) + \phi(2) + \phi(3) + \phi(4) + \phi(6) + \phi(12) \end{aligned}$$

Since all the $\phi(n)$'s together equal the number of residues, there must actually be $\phi(n)$ residues of each order n that is a factor of $p - 1$. Otherwise some residues would be left over. Especially, there must be $\phi(p - 1)$ residues of order $p - 1$. But these are primitive roots. The theorem is proved.

The Artist and the Astronaut

From the primitive-root theorem arose a disagreement among the admirers of the two great mathematicians. Gauss's fans say that he gave the first real proof, and Euler's fans say that *his* proof was perfectly good and that Gauss's objections to it were unimportant. It is quite clear, from what Euler wrote, that he had

Leonhard Euler, 1707–1783

in mind exactly the same arguments that went into Gauss's proof, and that he could have written it out correctly if he had tried. Still, he did not put the proof in a convincing form. He left himself in the position of a lawyer who may lose his case because he presented his arguments in faulty order, even though the arguments themselves are strong enough to win. Euler's proof is not even contained all in one piece of writing. Those who claim it was a complete proof have to piece it together out of remarks he made in different publications. Certainly, Gauss would never have published any proof in such disorganized form.

Before concluding that Euler was only a sloppy amateur, bear in mind that his aim in number theory was not to prove, so much as to discover. He wanted to know what was true, and when he found a new fact he wanted to share it with his readers. Naturally, he understood what a proof is, and was the first to find proofs of many statements which Fermat had bequeathed to the world as discoveries. But in his day no one expected, in a proof about numbers, the meticulous, lawyerlike care that we now call "mathematical rigor." This rigor had always been considered a part of geometry more than of other branches of mathematics.

Gauss set a new fashion in number theory by writing his *Disquisitions* in the rigorous manner of Euclid's *Elements.* It was in Gauss's nature never to be satisfied with anything imperfect. He couldn't bear to publish a discovery, even though everyone was eager to know it, unless he could present it in a way that satisfied him completely—and one of his requirements was a rigorous proof.

Euler had no such inhibitions. His papers give a feeling of spontaneity, of work in progress. It is as though he were saying, "This is all I could discover up to now, but if I think of anything new tonight, I will write another paper tomorrow." He would often publish when a problem was half-solved, then return to the subject years later with further progress. As a result, his discoveries in number theory are not contained in an orderly book like Gauss's *Disquisitions,* but are scattered through papers published over a period of forty years.

Gauss's writings give the opposite impression. Not only has he solved a problem completely before he writes about it; indeed, the reader feels that he has gone far beyond it, and that what he tells is only a small fraction of what he knows. This impression is quite correct. Gauss did know much more than he ever told. He kept quiet, not, like the Pythagoreans, for religious reasons, and not, like Fermat perhaps, to tease his colleagues, but simply because it took him a great deal of time to work anything into form. As a result, he published very little for such a great scholar. In his later years, friends often urged him to write more, even if it was imperfect. Once his ideas were known, other men could improve their form, but if he died without publishing, the ideas would be lost. These suggestions seemed to irritate him. He insisted that the extra care he put into his work had to do with the innermost part and not with finishing touches that anyone could add. To him, mathematics was an art like painting or sculpture. A painter does not write about how he imagines a painting might look. Either he produces the finished painting, or he forgets about it. But Euler was more like explorers and astronauts, who send out bulletins to keep the world informed.

A Problem in Two Parts

In Chapter 7 we studied Euler's work on the "hypotenuse" problem, that is, the problem of finding which numbers are sums of two squares. We also looked into what might be called the "generalized hypotenuse" problem, in which expressions like $x^2 + 3y^2$or $x^2 + 5y^2$ take the place of $x^2 + y^2$. Toward the end of the eighteenth century this generalized problem became quite celebrated and was a chief topic of interest for Lagrange, Legendre, and Gauss. (If you ever make a serious study of physics, you will have to keep separate places in your mind for Laplace, Laguerre, Lagrange, and Legendre. Only the last two, however, contributed to number theory.)

From Euler's work on $x^2 + y^2$ it was clear that this sort of problem has two parts. The first part is to find the relations between different numbers of the same form. For example, we saw that the product of any two primitive hypotenuses is a primitive hypotenuse (multiplication theorem), and that any factor of a primitive hypotenuse is a primitive hypotenuse (factor theorem). But for $x^2 + 5y^2$ only the multiplication theorem was true, not the factor theorem.

The second part is to find which numbers have multiples that obey the given form. This is a problem in residue theory. Thus, in the present chapter, we showed that every upper prime has a multiple that is a primitive hypotenuse. Or, in residue language, if p is prime and

$$p \equiv 1 \pmod 4$$

then some x exists such that

$$x^2 \equiv -1 \pmod p$$

Putting this together with the factor theorem, we concluded that every upper prime *is* a primitive hypotenuse.

It was Lagrange who found the key to the first part, for "generalized hypotenuses." He showed that it is not enough to study the forms $x^2 + cy^2$, or even $ax^2 + cy^2$. One must consider also forms $ax^2 + bxy + cy^2$. Thus, which numbers can be expressed as $3x^2 - xy - 2y^2$?

Expanding the problem in this way actually simplifies it. For one thing, certain forms are *equivalent* to each other. If we take the form $x^2 + 4xy - y^2$ and let $y = y' + x$, we find that

$$\begin{aligned} x^2 + 4xy - y^2 &= x^2 + 4xy' + 4x^2 - y'^2 - 2y'x - x^2 \\ &= 4x^2 + 2xy' - y'^2 \end{aligned}$$

Therefore the forms $x^2 + 4xy - y^2$ and $4x^2 + 2xy - y^2$ are equivalent. Any number (such as 19) which can be expressed in one form ($19 = 2^2 + 4 \times 2 \times 3 - 3^2$) can also be expressed in the other ($19 = 4 \times 2^2 + 2 \times 2 \times 1 - 1^2$) and *vice versa.*

Even if two forms are not equivalent, they may belong to the same *family.* For example, $x^2 + 5y^2$ and $2x^2 + 2xy + 3y^2$ belong to the same family of forms. The rule is as follows: if the form is $ax^2 + bxy + cy^2$, then the number $b^2 - 4ac$ is called the *discriminant.* All forms with the same discriminant belong to the same family. Incidentally, two equivalent forms always have the same discriminant.

The wonderful thing discovered by Lagrange is that the multiplication and factor theorems both apply to *families* of forms. Thus, $x^2 + 5y^2$ and $2x^2 + 2xy + 3y^2$, together with forms equivalent to one or the other, make up a family (discriminant -20). Therefore, if a number is of either of these forms, then all its factors will be of one form or the other. We have

$$42 = 2 \times 1^2 + 2 \times 1 \times (-4) + 3(-4)^2$$

and the factors of 42 are

$$\begin{aligned} 2 &= 2 \times 1^2 + 2 \times 1 \times 0 + 3 \times 0^2 \\ 3 &= 2 \times 0^2 + 2 \times 0 \times 1 + 3 \times 1^2 \\ 6 &= 1^2 + 5 \times 1^2 \\ 7 &= 2 \times 1^2 + 2 \times 1 \times 1 + 3 \times 1^2 \\ 14 &= 3^2 + 5 \times 1^2 \\ 21 &= 4^2 + 5 \times 1^2 \end{aligned}$$

Again, if two numbers of either form are multiplied, the product will be of one form or the other. Thus

$$3 \times 6 = 18 = 2 \times 3^2 + 2 \times 3 \times 0 + 3 \times 0^2$$
$$2 \times 42 = 84 = 8^2 + 5 \times 2^2$$

What made the form $x^2 + y^2$ so easy to deal with is that all the forms in its family are equivalent. Thus the factor theorem can be stated for it without involving any other forms.

Lagrange's discovery opened the way, but there remained many questions of detail which were studied by Legendre and Gauss. The central portion of Gauss's *Disquisitions* is concerned with the relations among inequivalent forms belonging to the same family. How many are there, which numbers can be expressed by which ones, and which form is obtained by multiplying two given forms together?

Gauss's treatment of this subject shows, by its striking contrast with Euler's work, how far the idea of *generality* had progressed in half a century. For Euler, each new form was a new problem. Having solved the problem of $x^2 + y^2$, he took a fresh start with $x^2 + 3y^2$, and so on. He was in the position of a grade-school child who, having mastered the 7-times multiplication table, prepares for a new challenge in the 8-times table. But Gauss (having the researches of Euler and Lagrange behind him) took it for granted that the essence of the problem was to find the laws governing *all* forms. He used particular forms to illustrate the laws, but his examples were not chosen for simplicity. He tossed around complicated forms like $17x^2 - 24xy + 73y^2$, as casually as our grade-school child's older brother would practice his skills in multiplying 73 by 82. His discussion was so complete that it left virtually nothing more to be done.

Quadratic Residues

The second part of the "generalized-hypotenuse" problem

boils down to finding out which residues are squares. For example, suppose that

$$x^2 \equiv 3 \qquad (\text{mod } p)$$

Then if we let $y = 1$, we have $x^2 - 3y^2$ a multiple of p. Or, if $x^2 \equiv -3$, then $x^2 + 3y^2$ is a multiple of p.

On the other hand, suppose that p has some multiple $x^2 + 3y^2$, where y is not 1. Then if p is prime, let z be the inverse of y, modulo p. We then have

$$\begin{aligned} 0 &\equiv (x^2 + 3y^2)z^2 \qquad (\text{mod } p) \\ &\equiv x^2z^2 + 3 \qquad (\text{mod } p) \end{aligned}$$

and -3 is again congruent to a square.

A similar trick works for the form $2x^2 + 3y^2$. If a number of this form has a prime factor p, then

$$\begin{aligned} 0 &\equiv 2(2x^2 + 3y^2) \qquad (\text{mod } p) \\ &\equiv (2x)^2 + 6y^2 \qquad (\text{mod } p) \end{aligned}$$

We can then multiply by the inverse of y and find a square congruent to -6. Or if we start with a square congruent to -6, we can find two numbers x and y such that $2x^2 + 3y^2$ is divisible by p. (Try proving this.)

These arguments were pointed out by Lagrange, but it was Legendre who first, in 1785, made a general attack on the problem of finding a square congruent to a given number. He called it the *quadratic-residue* problem. If

$$x^2 \equiv q \qquad (\text{mod } p)$$

then q is a quadratic residue of p. Thus 1, 2, and 4 are quadratic residues of 7, but 3 is not, since no square is congruent to 3, modulo 7. Again, -1 is a quadratic residue of 5 but not of 7 or 3.

Legendre made up a neat symbol for this relation, which is called the *Legendre symbol* and is written $\left(\frac{q}{p}\right)$. The Legendre symbol is not a fraction. It stands for a number which is equal

to 0 if $q \equiv 0 \pmod{p}$

to 1 if q is a quadratic residue of p not congruent to 0

to -1 if q is not a quadratic residue of p

You will see later why these rules are useful.

Thus, to say that a is a quadratic residue of b (but not divisible by it), we can just say

$$\left(\frac{a}{b}\right) = 1$$

Or, if we know that either a is a quadratic residue of b, or b of a, but not both, we can express our knowledge as

$$\left(\frac{a}{b}\right)\left(\frac{b}{a}\right) = -1$$

The quadratic-residue problem now becomes that of finding the value of $\left(\frac{a}{b}\right)$, for any a, b.

Legendre's Criterion

Legendre's first attempt in this direction was based on Lagrange's and Fermat's theorems. If p is an odd prime, and $\left(\frac{q}{p}\right) = 1$, then there is some residue x, other than zero, such that $x^2 \equiv q$, modulo p. By Fermat's Theorem $x^{p-1} \equiv 1$ or

$$q^{(p-1)/2} \equiv 1 \pmod{p}$$

(Since p is odd, ½$(p - 1)$ is a whole number.)

Now suppose that $\left(\frac{q}{p}\right) = -1$. By applying Fermat's Theorem to q, and noting that q^{p-1} is the square of $q^{(p-1)/2}$, we see that $q^{(p-1)/2}$ must be congruent either to 1 or to -1. But which is it? Lagrange's Theorem to the rescue!

Besides zero, there are $p - 1$ distinct residues. No more than two of these can have the same square (Lagrange's Theorem). Therefore there are ½(p − 1) distinct nonzero quadratic residues (i.e., squares). Each of these satisfies the congruence $q^{(p-1)/2} \equiv 1$, as we have shown. But this congruence has degree ½(p − 1). Therefore no other (nonquadratic residue) values of q satisfy it (Lagrange's Theorem again). So all the other nonzero residues must satisfy $q^{(p-1)/2} \equiv -1$.

Thus we have shown that $q^{(p-1)/2}$ is congruent, modulo p

to 1 if $\left(\frac{q}{p}\right) = 1$

to -1 if $\left(\frac{q}{p}\right) = -1$

and obviously it is congruent

to 0 if $\left(\frac{q}{p}\right) = 0$

since then $q \equiv 0$ by definition.

Or, in short,

$$q^{(p-1)/2} \equiv \left(\frac{q}{p}\right) \qquad (\text{mod } p)$$

whenever p is an odd prime. This is known as Legendre's criterion, because "criterion" means a rule for testing the truth of some statement. (Thus Wilson's Theorem is a criterion for whether n is prime.)*

To apply Legendre's criterion, let us ask whether 5 is a quadratic residue of 13. Here $q = 5$, $p = 13$, and we have to evaluate $5^{(13-1)/2}$ or 5^6. This is not so hard as it looks, because 13

* This criterion had been essentially proved by Euler in 1758. Lagrange's Theorem was not known then, and Euler used a different trick. Legendre's name is attached to the criterion because he first stated it clearly.

is a modulus. We have

$$5^2 = 25 \equiv -1 \pmod{13}$$
$$5^6 = (5^2)^3 \equiv (-1)^3$$

or

$$5^6 \equiv -1 \pmod{13}$$

and so, by Legendre's criterion,

$$\left(\frac{5}{13}\right) = -1$$

Thus 5 is not a quadratic residue of 13. You may check this by testing the squares of the numbers from 1 to 6 (why only these?) to see if any is congruent to 5.

From Legendre's criterion we learn some interesting rules for products. Let us keep one prime modulus p, but consider different values of q. Obviously, if we multiply two quadratic residues together, we get a quadratic residue, just as the product of two squares is always a square. But here we can go further. Suppose that $q = q_1q_2$. Then by taking the $\frac{1}{2}(p - 1)$ power, we find

$$q^{(p-1)/2} = q_1^{(p-1)/2}\, q_2^{(p-1)/2}$$

or

$$\left(\frac{q}{p}\right) = \left(\frac{q_1}{p}\right)\left(\frac{q_2}{p}\right)$$

Therefore, the product of a quadratic residue by a nonresidue (a number other than a quadratic residue) is a nonresidue, and the product of two nonresidues is a quadratic residue.

Although Legendre's criterion only states a congruence, we have been able to derive an equality from it. Since the Legendre symbols must be equal to 1, 0, or -1, the only way in which $\left(\frac{q}{p}\right)$ and $\left(\frac{q_1}{p}\right)\left(\frac{q_2}{p}\right)$ can be congruent is by being equal.

The distribution of quadratic residues can be best understood in terms of a primitive root. If r is a primitive root of p, it is easy to see that the even powers of r are quadratic residues, and that the odd powers are not (since the square of any power of r is an even power). From this the product rules also follow.

Quadratic Reciprocity

We now understand how $\left(\frac{q}{p}\right)$ varies with q, for a fixed prime p. But what Legendre really wanted to know was how $\left(\frac{q}{p}\right)$ varies with p, for a fixed q. Remember that the question, whether a prime p has a multiple $x^2 + 3y^2$, boiled down to asking whether $\left(\frac{-3}{p}\right) = 1$.

Euler had studied this question and found some surprising regularities, just by trial. You may check these yourself, for a few values of p.

If p is an odd prime, then

$$\begin{aligned} \left(\frac{2}{p}\right) &= 1 && \text{if } p \equiv 1 \text{ or } 7 && \pmod 8 \\ &= -1 && \text{if } p \equiv 3 \text{ or } 5 && \pmod 8 \end{aligned}$$

$$\begin{aligned} \left(\frac{3}{p}\right) &= 1 && \text{if } p \equiv 1 \text{ or } 11 && \pmod{12} \\ &= -1 && \text{if } p \equiv 5 \text{ or } 7 && \pmod{12} \end{aligned}$$

$$\begin{aligned} \left(\frac{5}{p}\right) &= 1 && \text{if } p \equiv 1 \text{ or } 4 && \pmod 5 \\ &= -1 && \text{if } p \equiv 2 \text{ or } 3 && \pmod 5 \end{aligned}$$

The last two lines are particularly interesting. They amount to the statement that

$$\left(\frac{5}{p}\right) = \left(\frac{p}{5}\right)$$

for all odd primes p.

The rule for $\left(\frac{3}{p}\right)$ seems less elegant. But after all, the statement "$p \equiv 7 \pmod{12}$" is just a short way of saying "$p \equiv 1 \pmod{3}$ and $p \equiv -1 \pmod{4}$." We already know that

$$\left(\frac{-1}{p}\right) = 1 \qquad \text{if } p \equiv 1 \pmod{4}$$
$$= -1 \qquad \text{if } p \equiv -1 \pmod{4}$$

By examining all the cases, you may see that the rule for $\left(\frac{3}{p}\right)$ can be restated:

$$\left(\frac{3}{p}\right) = 1 \qquad \text{if } \left(\frac{p}{3}\right) = \left(\frac{-1}{p}\right)$$
$$= -1 \qquad \text{if } \left(\frac{p}{3}\right) = -\left(\frac{-1}{p}\right)$$

or
$$\left(\frac{3}{p}\right)\left(\frac{-1}{p}\right) = \left(\frac{p}{3}\right)$$

or
$$\left(\frac{-3}{p}\right) = \left(\frac{p}{3}\right)$$

(Check this rule also.)

Legendre found that all the patterns discovered by Euler could be expressed in this way. Thus

$$\left(\frac{-7}{p}\right) = \left(\frac{p}{7}\right)$$
$$\left(\frac{-11}{p}\right) = \left(\frac{p}{11}\right)$$
$$\left(\frac{13}{p}\right) = \left(\frac{p}{13}\right)$$
$$\left(\frac{17}{p}\right) = \left(\frac{p}{17}\right)$$

for any odd prime p.

The minus sign is needed when the number other than p is a lower number. The general law is

$$\left(\frac{q}{p}\right) = \left(\frac{p}{q}\right) \qquad \text{if } q \text{ is an upper number}$$

$$\left(\frac{-q}{p}\right) = \left(\frac{p}{q}\right) \qquad \text{if } q \text{ is a lower number}$$

Note, however, that

$$\left(\frac{-q}{p}\right) = \left(\frac{-1}{p}\right)\left(\frac{q}{p}\right)$$

$$= \left(\frac{q}{p}\right) \qquad \text{if } p \text{ is an upper number}$$

$$= -\left(\frac{q}{p}\right) \qquad \text{if } p \text{ is a lower number}$$

So we are led to a beautiful statement, called the Law of Quadratic Reciprocity:

$$\left(\frac{q}{p}\right) = -\left(\frac{p}{q}\right) \qquad \text{if } p \text{ and } q \text{ are both lower numbers}$$

$$\left(\frac{q}{p}\right) = \left(\frac{p}{q}\right) \qquad \text{otherwise}$$

It is assumed that p and q are both odd primes.

10.

Proofs of a pudding

A Loophole

Legendre was the first to state the Law of Quadratic Reciprocity, although Euler stated parts of it. He went far beyond Euler, however, in attempting to prove it. Euler had been able to prove the rule $\left(\frac{-3}{p}\right) = \left(\frac{p}{3}\right)$ for all p. Legendre, in 1785, submitted a proof for all p and all q.

Legendre's proof was a by-product of a special interest of his, that of quadratic forms in *three* variables, such as $5x^2 - 2y^2 + 3z^2$. He solved many problems concerning such forms. For example, he proved that every odd number, except those congruent to 7, modulo 8, is the sum of three squares. Another important problem is this: If three prime numbers p, q, r are given, can one find three other numbers x, y, z so that

$$px^2 + qy^2 + rz^2 = 0$$

(Naturally p, q, r are allowed to be negative, since otherwise the answer would be no.)

By the arguments of Lagrange, it is easy to see that the equation cannot be satisfied unless

$$\left(\frac{-pq}{r}\right) = \left(\frac{-qr}{p}\right) = \left(\frac{-rp}{q}\right) = 1$$

(For example, take r as modulus, multiply the equation by p and by the inverse of y^2. One finds $(pxy^{-1})^2 \equiv -pq$, modulo r.)

Legendre was able to prove the converse, that if $\left(\frac{-pq}{r}\right)$, $\left(\frac{-qr}{p}\right)$, $\left(\frac{-rp}{q}\right)$ are all 1, and p, q, r are not all positive or all negative, then the desired x, y, z do exist. Let us accept this statement and see how it bears on quadratic reciprocity.

If $r = -1$, we can drop the requirement on $\left(\frac{-pq}{r}\right)$ since all numbers are congruent to one another with 1 as modulus. Then what Legendre has proved is that if $\left(\frac{p}{q}\right) = \left(\frac{q}{p}\right) = 1$, then x, y, z can be found so that

$$px^2 + qy^2 = z^2$$

But this is impossible if p and q are both lower numbers. For if x and y are both even, we can divide all three x, y, z by 2 until at least one (let us say x) is odd. Then x^2 will have to be an upper number, and px^2 a lower number. Take 4 as modulus; then $px^2 \equiv -1$, $qy^2 \equiv 0$ or -1 according as y is even or odd. Therefore $px^2 + qy^2$ is congruent to -1 or -2 and cannot be equal to z^2, which is congruent to 0 or 1.

The conclusion is that $\left(\frac{p}{q}\right) = \left(\frac{q}{p}\right) = 1$ is impossible when p, q are both lower primes. That is, of course, part of the Law of Quadratic Reciprocity. But it is only a part. The whole law has several cases, according as p and q are upper or lower and $\left(\frac{p}{q}\right) = 1$ or -1. (See if you can show that $\left(\frac{p}{q}\right) = \left(\frac{-q}{p}\right) = 1$ is impossible when p is lower and q is upper.) To prove all the cases, Legendre was forced to imagine suitable values of r. Thus he might need r to be an upper prime such that $\left(\frac{r}{p}\right) = 1$ and $\left(\frac{q}{r}\right) = -1$.

Legendre's proof was accepted as convincing by most mathematicians, and he is generally credited with discovering the law. However, Gauss pointed out that there was no guarantee that a suitable value of r would always exist. Perhaps there was some pair p, q for which no r would have the required properties. Therefore Legendre's proof was incomplete, and Gauss looked for a new one.

Four Lumps

Gauss based his proof on Euler's methods, which we studied in Chapter 7. The general idea is as follows. Suppose that q is an upper prime, and that $\left(\frac{q}{p}\right) = 1$. Then for some x, $x^2 - q$ is divisible by p. Now the reciprocity law says that $\left(\frac{p}{q}\right) = 1$. It says, in other words, that every odd factor of $x^2 - q$ is a quadratic residue of q. To prove this, we call the quadratic residues of q "good" and other numbers "bad." The number $x^2 - q$ is good since it is congruent to x^2. This part of the reciprocity law will be proved if we can show

(1) that a good number cannot have exactly one bad factor, and

(2) that if p has a multiple equal to $x^2 - q$, it has another multiple $y^2 - q$, less (in absolute value) than p^2.

Four difficulties arise in this proof. First, the theorem applies only to odd factors. Therefore, if 2 is a bad factor, it can spoil the proof for the others. This can be prevented by always making $x^2 - q$ an odd number. We shall want x less than p so that $x^2 - q$ will be less than p^2; but if x is odd we can replace it by $p - x$, which is even. Thus we obtain a value of x for which $x^2 - q$ is odd and less than p^2.

Second, we can prove (1) by Legendre's criterion, if $\left(\frac{x^2 - q}{q}\right) = 1$. For 1 cannot be the product of one factor -1 with any number of factors 1. But this proof breaks down if x is divisible by q. Then $\left(\frac{x^2 - q}{q}\right) = 0$, and 0 may be the product of -1 with 0.

This difficulty cannot be avoided by choosing x, since we have already used our choice in making x even. But we can remove the factor q: if $x = uq$ then $x^2 - q = q(u^2q - 1)$. The factors of $u^2q - 1$ are what we want, and $u^2q - 1$ is good if q is upper, since $\left(\frac{-1}{q}\right) = 1$.

If q is lower, the theorem deals with $\left(\frac{-q}{p}\right)$ and therefore with $x^2 + q$ or $q(u^2q + 1)$. Then $u^2q + 1$ is good.

Third, we intend to prove (2) by taking y congruent to x or to $-x$, and less than p. But this works only if q is small enough. If q is much bigger than p, $y^2 - q$ may be a negative number bigger than p^2, even though y is small. Trouble arises even more easily in the cases that deal with $y^2 + q$.

Fortunately, the proof of (2) is all right if q is less than p. Therefore if q is greater than p, we need only exchange the two numbers. The Law of Reciprocity is unchanged.

This makes the proof very intricate. Remember that a separate proof must be given for each part of the reciprocity law. The method of proof laid down in Chapter 5 is that starting with p and q which violate some part of the law, we use (1) to replace p by a smaller number, and then use (2) to replace x, and so on, generating an infinite descending sequence of violations. But now we find that if p has been replaced by a number less than q, we cannot apply (2) but must exchange p for q. Now q is smaller than p, and we may continue with (2). However, we now have a violation of a different part of the reciprocity law. Therefore the various parts cannot be proved independently, but all eight proofs must be interwoven by the induction reasoning.

The fourth difficulty is the most serious. This proof, like that of Legendre, deals more easily with negative statements than positive. Thus it shows that "bad" primes have *no* multiples of a certain form. But some parts of the law require us to show that every "good" prime *does* have some multiple of the form $x^2 - q$, or $x^2 + q$. This is a positive statement for which the method is not suitable.

Sometimes a positive statement about q can be converted into a negative one about $-q$. But this will not work when p is an upper prime. Nor are we allowed to convert a statement by exchanging p for q, since we have to keep q less than p to avoid the third difficulty. Hence Gauss was forced to the same extremity as Legendre: he converted a positive statement about q to a negative one about qr, where r is a third number satisfying a certain condition.

After all this work, Gauss was better off than Legendre in only one respect: his condition on r was simpler. In fact, he needed only to prove that for any prime p, there is a prime r less than p for which $\left(\frac{p}{r}\right) = -1$. It is evidence of his painstaking nature that he worked a whole year on this one proof. The proof is not interesting in itself, but it was most important to Gauss, for without it his demonstration of reciprocity would have been no better than Legendre's. As it was, he now possessed the first true, complete proof of the Law of Quadratic Reciprocity.

Taste

Mathematicians are hard to please. No sooner had Gauss

found this proof than he was looking for another one. The first one had left him unsatisfied.

He did not suspect any flaw in the proof. Intricate though it was, he had designed it carefully and knew that it was correct. Nor were there any gaps, unproven assumptions like those of Legendre. If the purpose of a proof is to establish the truth of a theorem, this one had done its job. He was convinced, but not satisfied.

In Chapter 1, I compared a proof to a dish of food and a theorem to a menu listing. Sometimes a dish is unsatisfactory because it doesn't have the right ingredients. The menu said beef, but the burger is of horsemeat. The juice isn't fresh and lacks Vitamin C. A dish that so fails to nourish is like a false proof, one that does not establish the truth of the theorem.

But a dish may nourish as advertised and still be awful. The burger is all beef, but it is cold and there is no salt. The sandwich is falling apart through holes in the bread. The juice has lumps, and the salad dressing has too much vinegar. Such a meal will keep us strong and healthy, but not happy. We crave dishes that look good, that taste good, that have just the right amount of everything.

Just so, a proof should do more than convince. It should illuminate. It should show why the theorem is true, not just that it is true. So, even after he had proved the reciprocity law, Gauss kept looking for shorter proofs, simpler proofs, more elegant, more illuminating proofs.

Disquisitiones Arithmeticae contains two proofs of reciprocity, the first one based, as shown above, on the methods of Chapter 7 and a second proof which Gauss found incidentally while studying the relations among quadratic forms. In the course of his life he is said to have discovered six additional proofs (two unpublished until after his death) but they are all variations on two main lines of reasoning. Each of these lines, most elegant and pleasing, is completely different from the other and from the two proofs in *Disquisitions*.

Gauss's Lemma

One line of proof is a fine example of the counting methods that have already shed such light on residue arithmetic. It is a variation of the argument I gave earlier, using Wilson's Theorem to show that $(2n)!^2 + 1$ is divisible by $4n + 1$, if $4n + 1$ is prime. In that argument I pointed out that the product of all the negative integers from -1 to $-a$ is just $(-1)^a$ times $a!$ From this it followed (supposing that $2a + 1 = p$, a prime modulus) that -1 would be a quadratic residue of p if $(-1)^a = 1$ (i.e., if p is an upper prime). Of course this is just Legendre's criterion applied to the residue -1. Now that we have proved the primitive-root theorem, we know that Legendre's criterion holds not only for -1 but for all residues. Therefore, instead of starting with Wilson's Theorem and ending with Legendre's criterion, we can run the argument backward, starting with Legendre's criterion and ending with new information.

Still supposing that $p = 2a + 1$, let q be any residue other than zero. If we multiply each of the residues $1, 2, \ldots, a$ by q and multiply all the results together, we have $q \times 2q \times 3q \times \ldots aq$, and this is obviously equal to $q^a a!$. Therefore, applying Legendre's criterion, we have

$$q \times 2q \times 3q \times \cdots \times aq \equiv \left(\frac{q}{p}\right) a! \qquad (\text{mod } p)$$

But there is another way of analyzing the product $q \times \cdots \times aq$. Let me call all the residues from 1 to a "positive residues" (modulo p) and the residues from $a + 1$ to $2a$ "negative residues" (since they are the same as $-a, \ldots, -1$). Thus, if $p = 5$, then 1 and 2 would be "positive," 3 and 4 "negative," 6 and 7 "positive" (same as 1 and 2), and so on.

Now if I write the multiples of q in order, $q, 2q, 3q, \ldots, 2aq$ (stopping not at aq but at $2aq$), I have a list which contains each nonzero residue just once. This follows from the fact that no residue is repeated, as you should be able to prove. (Note that I am using *multiples* of q; if I used *powers* of q, this statement would be true only if q were a primitive root.)

In the list of q-multiples, the last is the negative of the first, for

$$\begin{aligned} 2aq = (p-1)q &= pq - q \\ &\equiv -q \qquad \pmod{p} \end{aligned}$$

Likewise the next-to-last multiple is the negative of the second, and so on. This means that if the list is divided into two halves, each residue in the first half will have its negative in the last half, and *vice versa.* Hence the first half of the list is congruent to the numbers from 1 to a, but in a different order *and with some of the signs changed.*

For example, if $p = 7$ and $q = 2$, the first half of the list contains 2, 4, 6 which are congruent to 2, -3, -1. Or if $q = 3$, the first half of the list contains 3, 6, 9 which are congruent to 3, -1, 2. In each case there appear all the numbers from 1 to a ($=3$), but in a different order and with a different number of sign changes. It follows that

$$q \times 2q \ldots \times aq \equiv (-1)^{k_{qp}} a! \qquad \pmod{p}$$

where k_{qp} is the number of sign changes; that is, k_{qp} is the number of "positive" residues of p which, when multiplied by q, yield "negative" residues.

Comparing this congruence with the one derived from Legendre's criterion, we have

$$\left(\frac{q}{p}\right) \equiv (-1)^{k_{qp}} \qquad \pmod{p}$$

or

$$\left(\frac{q}{p}\right) = (-1)^{k_{qp}}$$

since both sides can only be 1 or -1.

This fact is called "Gauss's Lemma" and leads to a full knowledge about quadratic residues, if we can only get information about k_{qp}.

In The Bag

When $q = 2$, k_{qp} is found easily. It is just the number of integers greater than ½a, but not greater than a. Clearly then,

$$k_{2p} = \tfrac{1}{2}a = \tfrac{1}{4}(p-1) \qquad \text{if } a \text{ is even}$$
$$k_{2p} = \tfrac{1}{2}(a+1) = \tfrac{1}{4}(p+1) \qquad \text{if } a \text{ is odd}$$

Now if $p - 1$ or $p + 1$ is divisible by 8, then k_{2p} is even and $\left(\frac{2}{p}\right) = 1$. But otherwise k_{2p} is odd, and then $\left(\frac{2}{p}\right) = -1$. In other words, 2 is a quadratic residue of p if $p \equiv 1$ or 7, modulo 8, but not if $p \equiv 3$ or 5, modulo 8.

When q is an odd prime, k_{qp} is not so easy to calculate. The important thing is to relate k_{qp} to k_{pq}. Since we already know the *statement* of quadratic reciprocity, we can say in advance what we hope to prove. Let $p = 2a + 1$, and $q = 2b + 1$. Then, if p and q are both lower primes, a and b will both be odd, and so ab will be odd. Otherwise, at least one of a and b will be even, and ab will be even. Thus quadratic reciprocity says that $\left(\frac{p}{q}\right)\times\left(\frac{q}{p}\right)=1$ if ab is even, -1 if ab is odd. That is,

$$\left(\frac{p}{q}\right) \times \left(\frac{q}{p}\right) = (-1)^{ab}$$

But from Gauss's Lemma we know that

$$\left(\frac{p}{q}\right) \times \left(\frac{q}{p}\right) = (-1)^{k_{pq}}(-1)^{k_{qp}}$$

Therefore, to prove quadratic reciprocity, we need only prove that

$$(-1)^{ab} = (-1)^{k_{pq}+k_{qp}}$$

or, in other words, that

$$ab \equiv k_{pq} + k_{qp} \qquad (\text{mod } 2)$$

Now, consider all the numbers from 1 to just less than $\frac{1}{2}pq$. I shall call this set of numbers "the bag." According to principles already explained, each number in the bag is either divisible by p, or "positive" or "negative" with respect to p. Likewise it is either divisible, positive, or negative with respect to q. For example, if $p = 3$ and $q = 5$, then the bag holds the numbers 1 to 7. The character of each number is shown in the table.

With respect to		3	5
The number	is		
1		positive	positive
2		negative	positive
3		divisible	negative
4		positive	negative
5		negative	divisible
6		divisible	positive
7		positive	positive

From the original definition of k_{qp} we see that there are just k_{qp} numbers in the bag that are divisible by q and negative with respect to p. Likewise there are k_{pq} numbers in the bag, divisible by p and negative with respect to q. Let us also define numbers for the other categories. Say that there are n_{++} numbers in the bag, positive with respect to both p and q; n_{--} numbers, negative with respect to both; n_{+-} numbers, positive with respect to p but negative with respect to q; and n_{-+} numbers, negative to p and positive to q.

Altogether, then, the bag holds $n_{-+} + n_{--} + k_{qp}$ numbers that are negative residues of p. But you may easily see that there are just ab such numbers (they come in b clumps of a numbers each). So

$$n_{-+} + n_{--} + k_{qp} = ab$$

Likewise the bag holds

$$n_{+-} + n_{--} + k_{pq} = ab$$

negative residues of q. (They come in a clumps of b numbers each.) Check this on the table, noting that $a = 1$ and $b = 2$.

Now, consider all numbers from 1 to pq, both in and out of the bag, that are positive to p and negative to q. Each such number which is *not* in the bag is congruent, modulo pq, to the negative of a number *in* the bag, which is negative to p and positive to q. Therefore there are altogether $n_{+-} + n_{-+}$ numbers from 1 to pq that are positive to p and negative to q. But, on the other hand, there are exactly ab such numbers. For there are a positive residues of p, and b negative residues of q, and each combination is represented just once by a number from 1 to pq. (Check this also for $p = 3$, $q = 5$.) Hence

$$n_{+-} + n_{-+} = ab$$

If we add the two equations containing k_{pq} and k_{qp}, and subtract this last, the n_{+-} and n_{-+} cancel out and we have

$$\begin{aligned} 2n_{--} + k_{pq} + k_{qp} &= 2ab - ab \\ &= ab \end{aligned}$$

or

$$k_{pq} + k_{qp} \equiv ab \pmod 2$$

which proves the theorem.

Wizardry

The other line of proof is my favorite because it reveals the touch of mystery always latent in mathematics. It starts from a special trick that was already known to Euler, a trick which shows that every prime of the form $p = 3n + 1$ has a multiple of the form $a^2 + 3$, or, in other words, that -3 is a quadratic residue of p, if p is a prime quadratic residue of 3.

The trick depends on the existence of primitive roots. Let r be a primitive root of $p = 3n + 1$, and let $x \equiv r^n$. Then $x \not\equiv 1$, but $x^3 \equiv 1$, modulo p by Fermat's Theorem, applied to r. In other words, $x^3 - 1$, or $(x - 1)(x^2 + x + 1)$, is divisible by p but $x - 1$ is not. Therefore

$$x^2 + x + 1 \equiv 0 \pmod p$$

From this it follows that the square of $(x - x^2)$ is congruent to -3, so that -3 is a quadratic residue of p. For

$$\begin{aligned}(x - x^2)^2 &= x^2 - 2x^3 + x^4 \\ &\equiv x^2 - 2 + x \pmod p \\ &\equiv (x^2 + x + 1) - 3 \equiv -3 \pmod p\end{aligned}$$

Euler and Lagrange, wizards of algebra, extended this trick to more difficult cases. If $p = 5n + 1$ and $x \equiv r^n$ where r is a primitive root, then

$$x^4 + x^3 + x^2 + x + 1 \equiv 0 \pmod p$$

It follows that 5 is congruent to the square of $(x - x^2 - x^3 + x^4)$. For

$$
\begin{aligned}
(x - x^2 - x^3 + x^4)^2 &= x^2 + x^4 + x^6 + x^8 - 2xx^2 - 2xx^3 + 2xx^4 \\
&\qquad + 2x^2x^3 - 2x^2x^4 - 2x^3x^4 \\
&\equiv x^2 + x^4 + x + x^3 - 2x^3 - 2x^4 + 2 \\
&\qquad + 2 - 2x - 2x^2 \\
&= 4 + (x + x^2 + x^3 + x^4) - 2(x + x^2 \\
&\qquad + x^3 + x^4) \\
&= 5 - (1 + x + x^2 + x^3 + x^4) \\
&\equiv 5 \pmod{p}
\end{aligned}
$$

Again, if $p = 7n + 1$, we can find x so that

$$1 + x + x^2 + x^3 + x^4 + x^5 + x^6 \equiv 0 \pmod{p}$$

Then it follows that

$$(x + x^2 - x^3 + x^4 - x^5 - x^6)^2 \equiv -7 \pmod{p}$$

(Work this one out for yourself.)

Try an example. Let $n = 2$, $p = 11 = 5n + 1$. Let $r = 2$, which happens to be a primitive root of 11. Then $x \equiv r^n = 2^2 = 4$. Check that $1 + x + x^2 + x^3 + x^4$ is divisible by 11. Then form the quantity $x - x^2 - x^3 + x^4 \equiv 4 - 16 - 64 + 256 = 180$. Sure enough, $180^2 \equiv 4^2 = 16 \equiv 5$, modulo 11.

Science

Lagrange carried this trick only as far as 7. But Gauss discovered the principle behind it and showed in *Disquisitions* how, if q is any odd prime and $p = qn + 1$ is also prime, one can construct a number whose square is congruent, modulo p, to q or $-q$, according as q is an upper or a lower number. First, let r be a primitive root of p, and $x \equiv r^n$, as before, so that

$$1 + x + x^2 + \cdots + x^{q-1} \equiv 0 \pmod{p}$$

Second, add together all the powers of x from the first to the

$(q - 1)$st, putting a plus sign before any power if its exponent is a quadratic residue of q, and a minus sign otherwise. I shall call this expression $f_q(x)$, or simply f_q. Thus

$$f_5(x) = f_5 = x - x^2 - x^3 + x^4$$

where the first and fourth powers are taken with a plus sign, since 1 and 4 are quadratic residues of 5, and the second and third powers have a minus sign since 2 and 3 are not quadratic residues. Again, the quadratic residues of 11 are 1, 3, 4, 5, and 9; therefore

$$f_{11} = x - x^2 + x^3 + x^4 + x^5 - x^6 - x^7 - x^8 + x^9 - x^{10}$$

In general, we can write f_q with the help of Legendre symbols

$$f_q(x) = \left(\frac{1}{q}\right)x + \left(\frac{2}{q}\right)x^2 + \cdots + \left(\frac{q-1}{q}\right)x^{q-1}$$

The idea of using quadratic residues in this way was one of Gauss's finest discoveries in this field, and one in which he went quite beyond Euler and Lagrange.

To evaluate $f_q{}^2$ in general, we cannot multiply straight out because f_q is a different expression for each value of q. So we proceed indirectly. What happens to f_q if we replace x by some power of x? In other words, what is the relation between $f_q(x^s)$ and $f_q(x)$?

Suppose, for example, that $q = 5$, and replace x by x^2. We have

$$\begin{aligned} f_5(x^2) &= x^2 - (x^2)^2 - (x^2)^3 + (x^2)^4 \\ &= x^2 - x^4 - x^6 + x^8 \end{aligned}$$

But since x has been chosen so that $x^5 \equiv 1$, modulo p, we can reduce the exponents and write

$$f_5(x^2) \equiv x^2 - x^4 - x + x^3 \qquad (\text{mod } p)$$

$$= -x + x^2 + x^3 - x^4 = -f_5(x)$$

In the same way

$$\begin{aligned} f_5(x^4) &= x^4 - x^8 - x^{12} + x^{16} \\ &\equiv x^4 - x^3 - x^2 + x \qquad \pmod p \\ &= f_5(x) \end{aligned}$$

It is easy to see a principle behind these examples. When x is replaced by x^2, all the exponents in f_5 are multiplied by 2. Since 2 is not a quadratic residue of 5, this turns all the quadratic residues into nonquadratic residues and vice versa. Hence the negative terms in $f_5(x^2)$ have quadratic-residue exponents and the positive terms do not. This makes $f_5(x^2)$ congruent modulo p to $-f_5(x)$. But when x is replaced by x^4, the exponents are multiplied by 4 and so the positive terms still have quadratic residue exponents.

Clearly, no matter what odd prime we use for q, as long as x is chosen so that $x^q \equiv 1$, modulo p, the same reasoning shows that $f_q(x^s) \equiv \left(\frac{s}{q}\right) f_q(x)$, modulo p. It follows (squaring both sides) that $f_q(x^s)^2 \equiv f_q(x)^2$, modulo p, as long as s is not divisible by q.

But this fact tells us a great deal about $f_q{}^2$. Certainly $f_q{}^2$ is congruent modulo p to some combination of terms

$$a_0 + a_1x + a_2x^2 + \cdots + a_{q-1}x^{q-1}$$

where a_0, etc., are numbers from 1 to p which we do not know. If this expression is to remain unchanged when x is replaced by $x^2, x^3, \ldots, x^{q-1}$, all the numbers $a_1, a_2, \ldots a_{q-1}$ must be the same, and hence

$$f_q{}^2 = a_0 + a_1(x + x^2 + \cdots + x^{q-1}) \qquad \pmod p$$

Now there are only two unknown numbers. To find a_0, observe that there are $q - 1$ products in which a term of f_q is multiplied by another term of f_q so as to get a term $x^q \equiv 1$. Each of these products is of the form

$$\left(\frac{s}{q}\right)x^s \times \left(\frac{q-s}{q}\right)x^{q-s}, \text{ or } \left(\frac{s}{q}\right)\left(\frac{-s}{q}\right)x^q.$$

But
$$\left(\frac{s}{q}\right)\left(\frac{-s}{q}\right) = \left(\frac{-s^2}{q}\right) = \left(\frac{-1}{q}\right)$$

Hence all these products have the same sign, which is positive if q is upper, negative if q is lower. It follows that

$$a_0 = \left(\frac{-1}{q}\right)(q-1)$$

To find a_1, we can substitute the value 1 for x. This is because everything we have done so far has depended only on the congruence $x^q \equiv 1$, which is true if $x \equiv 1$, and we have not yet used the stronger restriction $1 + x + \cdots + x^{q-1} \equiv 0$, which excludes 1. If $x \equiv 1$, then all its powers are 1, and so

$$f_q(1)^2 \equiv a_0 + a_1(q-1) \qquad (\text{mod } p)$$

On the other hand, $f_q(1) \equiv 0$ since half the terms are 1 and half are -1. Therefore $f_q(1)^2 = 0$, and

$$a_1(q-1) \equiv -a_0 \qquad (\text{mod } p)$$

or
$$a_1 \equiv -\left(\frac{-1}{q}\right)$$

We now have

$$\begin{aligned} f_q^2 &\equiv \left(\frac{-1}{q}\right)(q-1) - \left(\frac{-1}{q}\right)(x + \cdots + x^{q-1}) \\ &= \left(\frac{-1}{q}\right)q - \left(\frac{-1}{q}\right)(1 + x + \cdots + x^{q-1}) \\ &\equiv \left(\frac{-1}{q}\right)q \qquad (\text{mod } p) \end{aligned}$$

where we have finally substituted a value of x for which $1 + x + \cdots + x^{q-1} \equiv 0$. This argument is one of the earliest examples (others are found in the work of Lagrange) of a method that has become extremely important in modern algebra, in which the properties of an expression are deduced, not by calculating it directly, but by showing that it remains unchanged under various substitutions or transformations.

Shadows

Gauss had worked all this out when he wrote *Disquisitions*, and he offered it then as a partial proof of quadratic reciprocity. It shows, in fact, that if $p \equiv 1$, modulo q, then $\pm q$ is a quadratic residue of p. (From now on I shall write $\pm q$ instead of $\left(\frac{-1}{q}\right)q$; thus $\pm$ means $+$ if q is upper, or $-$ if q is lower.) But $\left(\frac{-1}{q}\right)$ is a quadratic residue of p, unless p and q are both lower numbers; therefore we have proved that if $p \equiv 1$, modulo q then q is a quadratic residue of p unless q and p are both lower. This is just what quadratic reciprocity says, since p is obviously a quadratic residue of q in this case.

The remarkable thing is that Gauss was later able to extend this argument to a full proof in which he showed, even if $p - 1$ is not divisible by q, that $\left(\frac{\pm q}{p}\right) = \left(\frac{p}{q}\right)$, that is, $\pm q$ will be a quadratic residue of p if p is a quadratic residue of q, and not otherwise. It is remarkable because the whole argument depends on the congruence

$$1 + x + \cdots + x^{q-1} \equiv 0 \qquad (\text{mod } p)$$

and if $p - 1$ is not divisible by q, there is no residue x for which this congruence is true. Hence we cannot show that $\pm q$ is a quadratic residue of p by making it congruent to the square of $f_q(x)$.

Nevertheless, the special nature of f_q can be brought out regardless of the value of p. Its virtue is that *if* there were an x for which $1 + x + \cdots + x^{q-1} = 0$, then the square of $f_q(x)$ *would* be $\pm q$. At first sight such a statement seems foreign to mathematics, in which everything is either true or false. We do not say that $3 + 3$ *may* be 5, or that it *would* be if only 8 were less than 2. Yet this shadow of a truth can be made clear and definite by a new idea, that of using the expression $1 + x + x^2 + \cdots + x^{q-1}$ as a modulus.

In this new mode of thought, we do not regard x as a number or a residue at all. Instead we regard each expression $1 + x$, $2 - x^2 + x^3$, etc., as a "mathematical object" with certain formal qualities. These qualities belong to the expression just as it stands (without our giving any particular value to x). They are called "formal" because they depend on the form of the expression and not on its numerical value or on that of x.

We have already examined such properties in Chapter 8, where I pointed out that a polynomial may be identically (that is, formally) divisible by a number, as $3x^2 + 3x$ is by 3, since $3x^2 + 3x = 3(x^2 + x)$ regardless of the value of x. In the same way, a polynomial may be formally divisible by another polynomial.

For example, $2 - x^2 + x^3$ is *formally* a multiple of $1 + x$, since

$$2 - x^2 + x^3 = (1 + x)(2 - 2x + x^2)$$

This is a truth independent of the value of x. Another way of stating it is that

$$2 - x^2 + x^3 \equiv 0 \qquad (\text{mod } 1 + x)$$

In other words, two expressions are congruent modulo $1 + x$ if their difference is the *formal* product of $1 + x$ and some other expression.

Similarly

$$x^2 \equiv 1 \qquad (\text{mod } 1 + x)$$

since $1 - x^2$ is the formal product of $1 + x$ and $1 - x$. (Note that 1 is treated as an expression on the same footing as x^2.) The very same arguments as those used in Chapter 6 will show that these congruences can be added, subtracted, and multiplied like equations, but not divided, unless the modulus has no factors.

If the modulus is $1 + x + x^2 + \cdots + x^{q-1}$, then $x^q \equiv 1$ since $1 - x^q$ is the product of this modulus with $1 - x$. In fact, all the congruences that we derived before, under the assumption that $1 + x + x^2 + \cdots + x^{q-1} \equiv 0$, modulo p, can be derived now as congruences modulo $1 + x + \cdots + x^{q-1}$, without any assumption about x. The same arguments now lead to

$$f_q^2 \equiv \pm q \qquad (\text{mod } 1 + x + \cdots + x^{q-1})$$

This clear assertion expresses the shadowy notion we began with, that f_q^2 would be $\pm q$ *if* $1 + x + \cdots + x^{q-1}$ were 0. Mathematics is really full of shadowy impulses, and it is the work of creative mathematicians to spot them and transform them into exact truths. Thus mathematics demands the same two qualities as poetry, music, and the other arts. The mathematician must be sensitive, so that he feels the presence of an idea that has no definite form; and he must have an exact mind, so that he can fashion a form that fits the idea.

Skill

This new congruence on f_q^2 is true of any prime q, but it does not involve p as modulus and therefore says nothing about the relation of q to p. In order that $\pm q$ be a quadratic residue of p, there must be a *residue* of p whose square is $\pm q$. An *expres-*

sion like f_q will not suffice. Still, we may find out something by putting f_q^2 into Legendre's criterion

$$\left(\frac{\pm q}{p}\right) \equiv (\pm q)^{(p-1)/2} \qquad (\text{mod } p)$$

but

$$(\pm q)^{(p-1)/2} \equiv f_q^{p-1} \qquad (\text{mod } 1 + x + \cdots + x^{q-1})$$

and therefore

$$\left(\frac{\pm q}{p}\right) \equiv f_q^{p-1} \qquad (\text{mod } p,\ 1 + x + \cdots + x^{q-1})$$

The last congruence has *two* moduli, and its meaning is that the difference between $\left(\frac{\pm q}{p}\right)$ and f_q^{p-1} can be obtained by adding a formal multiple of p to a formal multiple of $1 + x + \ldots x^{q-1}$. As an example, take $q = 3$, $p = 5$. Then $\left(\frac{\pm q}{p}\right) = \left(\frac{-3}{5}\right) = \left(\frac{2}{5}\right) = -1$; and $f_q^{p-1} = (x - x^2)^4 = x^4 - 4x^5 + 6x^6 - 4x^7 + x^8$. The difference, $1 + x^4 - 4x^5 + 6x^6 - 4x^7 + x^8$, can be obtained by adding

$$-5x^5 + 5x^6 - 5x^7 = 5(-x^5 + x^6 - x^7)$$

to

$$1 + x^4 + x^5 + x^6 + x^7 + x^8 = (1 + x + x^2)(1 - x + x^3 + x^6)$$

Now you are thinking that we should use Fermat's Theorem and replace f_q^{p-1} by 1. But this is not allowed, for Fermat's Theorem applies to residues, not to polynomials. Remember that x^3 is not identically congruent to x modulo 3, nor $(2x + 1)^3$ to $2x + 1$.

However, there is a peculiar result of raising any polynomial to a prime power p, with p as modulus; we have already encountered it in one of Euler's proofs of Fermat's Theorem. We noted there that $(x + y)^p$ is congruent modulo p to $x^p + y^p$ since all the other terms, like $\frac{p(p-1)}{2}x^{p-2}y^2$, are divisible by p. Now observe that this congruence is formal, that it holds just as well if we start with many terms instead of two, and that the terms can themselves be powers of x. For example, $(2 + x - 5x^2)^3 \equiv 2^3 + x^3 + (-5x^2)^3$, modulo 3 as you can verify by multiplication. The remaining terms occur in triplets and can be dropped.

Generally, if

$$g(x) = b_0 + b_1x + \cdots + b_nx^n$$

is any polynomial, then

$$g(x)^p \equiv b_0{}^p + (b_1x)^p + \cdots + (b_nx^n)^p \qquad (\text{mod } p)$$

or (using Fermat's theorem on the numbers $b_0 \ldots b_n$)

$$g(x)^p \equiv b_0 + b_1x^p + \cdots + b_n(x^p)^n$$

or

$$g(x)^p \equiv g(x^p) \qquad (\text{mod } p)$$

This pretty congruence is the extension of Fermat's Theorem to polynomials.

Beauty

The extended version of Fermat's Theorem is especially useful for f_q, since we have already shown that

$$f_q(x^s) \equiv \left(\frac{s}{q}\right) f_q(x) \qquad (\text{mod } 1 + x + \cdots + x^{q-1})$$

for any s. (This was first stated with p as modulus, under the assumption that x was a *residue* of p for which $1 + x + \cdots + x^{q-1} \equiv 0$. But if we do not make such an assumption we must use $1 + x + \cdots + x^{q-1}$ as modulus.) Taking $s = p$, we now have

$$f_q(x)^p \equiv f_q(x^p) \qquad (\text{mod } p)$$

and $\quad f_q(x^p) \equiv \left(\dfrac{p}{q}\right) f_q(x) \qquad (\text{mod } 1 + x + \cdots + x^{q-1})$

or $\quad f_q(x)^p \equiv \left(\dfrac{p}{q}\right) f_q(x) \qquad (\text{mod } p,\ 1 + x + \cdots + x^{q-1})$

Here we have a second congruence with two moduli, but this one involves $\left(\dfrac{p}{q}\right)$ instead of $\left(\dfrac{\pm q}{p}\right)$. The earlier congruence (derived from Legendre's criterion) may be multiplied by $f_q(x)$, giving

$$\left(\frac{\pm q}{p}\right) f_q \equiv f_q^p \qquad (\text{mod } p,\ 1 + x + \cdots + x^{q-1})$$

and this may be combined with the new congruence (derived from the extended Fermat's Theorem) to give

$$\left(\frac{\pm q}{p}\right) f_q \equiv \left(\frac{p}{q}\right) f_q \qquad (\text{mod } p,\ 1 + x + \cdots + x^{q-1})$$

Shall we now divide by f_q to obtain the reciprocity law? Not allowed! Even though p and $1 + x + \cdots + x^{q-1}$ are prime when taken separately, $1 + x + \cdots + x^{q-1}$ may not be prime in the presence of the modulus p. Thus

$$1 + x + x^2 \equiv (1 - 2x)(1 + 3x) \qquad (\text{mod } 7)$$

We may write

$$(1 - 2x)(1 + 3x) \equiv 0 \qquad (\text{mod } 7,\ 1 + x + x^2)$$

but if we divide by $1 - 2x$ we obtain the false statement

$$1 + 3x \equiv 0 \qquad (\text{mod } 7,\ 1 + x + x^2)$$

Instead of dividing, let us multiply by f_q

$$\left(\frac{\pm q}{p}\right) f_q^2 \equiv \left(\frac{p}{q}\right) f_q^2 \qquad (\text{mod } p,\ 1 + x + \cdots + x^{q-1})$$

or since $f_q^2 \equiv \pm q$,

$$\pm q\left(\frac{\pm q}{p}\right) \equiv \pm q\left(\frac{p}{q}\right) \qquad (\text{mod } p,\ 1 + x + \cdots + x^{q-1})$$

This congruence has nothing to do with x, since f_q is eliminated. Therefore the modulus $1 + x + \cdots + x^{q-1}$ is unnecessary,

and
$$\pm q\left(\frac{\pm q}{p}\right) \equiv \pm q\left(\frac{p}{q}\right) \qquad (\text{mod } p)$$

Having disposed of one modulus, we can now divide by $\pm q$ since the remaining modulus is prime. This gives

$$\left(\frac{\pm q}{p}\right) \equiv \left(\frac{p}{q}\right) \qquad (\text{mod } p)$$

Finally, we can drop the modulus p because the Legendre symbols must be 1 or -1. We obtain at last the quadratic-reciprocity theorem for odd primes:

$$\left(\frac{\pm q}{p}\right) = \left(\frac{p}{q}\right)$$

If $q = 2$, this method does not work. But we can instead let $q = 8$, since $\left(\frac{8}{p}\right) = \left(\frac{2}{p}\right)$ for all p. We take $x^4 + 1$ as a second modulus (so that $x^8 \equiv 1$) and define

$$f_8 = x - x^3 - x^5 + x^7$$

It then turns out that $f_8{}^2 \equiv 8$, modulo $x^4 + 1$ and that $f_8(x^p) \equiv f_8(x)$ if $p - 1$ or $p + 1$ is a multiple of 8, or $-f_8(x)$ otherwise. Hence it follows as before (see page 200) that 2 is a quadratic residue of any prime $p = 8n + 1$ or $8n + 7$, but not of $p = 8n + 3$ or $8n + 5$.

Isn't that beautiful?

Further Reading

I. Mathematics for Pleasure

Lewis Carroll, *The Game of Logic,* N. Y., Dover (1896). An introduction to formal logic, truth tables, and syllogisms. Carroll is his irrepressible self, but quite serious about the material.

Tobias Dantzig, *Number, the Language of Science,* N. Y., Macmillan (1941). A pioneer among modern books on mathematics for the layman. Aimed at the broadly educated adult, it traces the successsive generalizations of number from positive to negative, whole to fractional, rational to irrational, algebraic to transcendental, real to complex, finite to transfinite. It emphasizes the historical development of concepts rather than anecdotes or technical details, both of which, however, are present.

Maurice Kraitchik, *Mathematical Recreations,* N. Y., Norton (1942). A rich trove of number games, puzzles, and curiosities.

George Gamow, *One, Two, Three . . . Infinity,* rev. ed., N. Y., Viking (1961). Covers a wide range of mathematical subjects, including numbers, the infinite, and topology. Written for the young at heart. (Reprinted by Dover, 1987).

Abraham Fraenkel, *Integers and Theory of Numbers,* N. Y., Scripta Mathematica (1955). Number theory and other mathematical topics. The material is elementary, but the treatment is profound and sophisticated.

Constance Reid, *From Zero to Infinity: What Makes Numbers Interesting,* N. Y., Crowell (1955). An excellent short book with roughly the material of our chapters 1–7. The author devotes a chapter to each number from 0 to 9.

Isaac Asimov, *Realm of Numbers,* Boston, Riverside Press (1959). Same scope as Dantzig, but shorter, more direct, more detailed in explanation, and less profound.

II. About Mathematicians

Eric Temple Bell, *Men of Mathematics,* N. Y., Simon and Schuster (1937). Interesting short biographies of the great mathematicans of history, with some discussion of their work.

James R. Newman, ed., *World of Mathematics,* N. Y., Simon and Schuster (1956). This four-volume collection of writings by and about mathematicans makes wonderful browsing.

Leonard E. Dickson, *History of the Theory of Numbers,* N. Y., Stechert (1934). For reference only. An exhaustive list of significant contributions to number theory up to the time of compilation.

III. Classics of Number Theory

Euclid, *Elements,* ed. by Thomas L. Heath. N. Y., Dover (1926). The first six books are on plane geometry. Number theory is treated in Books 7–9. The material there is on the level of our chapter 3, but unfortunately it is obscurely presented. Book 10, though difficult, is worth study. A great masterpiece, it deals with the irrationals generated by the taking of square roots in combination with the various operations of arithmetic.

T. L. Heath, *Diophantus of Alexandria,* 2nd rev. ed., N. Y., Dover (1910). This is not a biography. It consists of an English translation, very easy to read, of Diophantus' *Arithmetica,* preceded by a historical discussion and accompanied by footnotes which contain much related material due to Fermat, Euler, and others.

Carl Friedrich Gauss, *Disquisitiones Arithmeticae,* trans. A. A. Clarke, New Haven and London, Yale University Press (1966). The Master's voice. Until recently there was no English translation. This one is readable, but there are easier ways to learn the subject.

IV. More about Number Theory

G. B. Mathews, *Theory of Numbers,* Cambridge, Deighton Bell (1892). Begins with congruences and develops systematically the subject of quadratic forms, to which our chapters 8–10 furnish an introduction.

L. E. Dickson, *Introduction to the Theory of Numbers,* Chicago, University of Chicago Press (1929). Mostly on quadratic forms. Goes beyond this book, but not as far as Mathews.

G. L. Hardy and E. M. Wright, *An Introduction to the Theory of Numbers,* Oxford, Clarendon Press (1938). A large text on the same level as Mathews but with considerable emphasis on the behavior of very large numbers, a subject not dealt with in the present book.

J. V. Uspensky and M. A. Heaslet, *Elementary Number Theory,* N. Y. and London, McGraw-Hill (1939). A good sequel to read after this book. It covers much of our material, but more completely, and includes problems for the reader.

B. M. Stewart, *The Theory of Numbers,* N. Y., Macmillan (1952). A modern college text on the level of Dickson's *Introduction,* but with more variety of material.

Daniel Shanks, *Solved and Unsolved Problems in Number Theory,* Washington, Spartan Books (1962). An exposition of the subject which stresses the extent and limits of present knowledge.

Z. I. Borevich and I. R. Shafarevich, *Number Theory,* trans. N. Greenleaf, N. Y., Acad. Press (1966). Presents the subject as it has been transformed since Gauss with the notions of modern algebra.

Table of Theorems

All the theorems that appear in this book are listed below. They are grouped into categories A–L according to subject:

Category A:	Sums of series
Category B:	Congruences (theorems asserting that one number is congruent to another)
Category C:	Laws of congruences (theorems telling what can be deduced from a congruence)
Category D:	Impossibilities (theorems asserting that numbers with certain properties cannot exist)
Category E:	Prime numbers (theorems asserting that certain numbers are or are not prime)
Category F:	Factors (general theorems about the factors of any number)
Category G:	Factors (theorems about the factors of special kinds of numbers)
Category H:	Formulas (theorems telling how to build up certain kinds of number)
Category I:	Quadratic residues
Category J:	Powers of residues
Category K:	Quadratic forms (theorems about $x^2 + y^2$ and similar expressions)
Category L:	Conjectures and false "theorems"

LABEL	THEOREM	REFERENCE
A1	Sum of first n cubes	Stated p. 19, p. 39 Proof indicated p. 39, details left to reader
A2	Sum of first n odd numbers	Stated p. 31 with proof
A3	Sum of first n numbers	Stated p. 32 with proof
A4	Sum of first n squares	Stated pp. 33–4, proof pp. 37–8
A5	Sum of first n powers of 2	Stated pp. 38–9 with proof
B1	Any odd square $\equiv 1 \pmod 8$	Stated p. 19, proof pp. 127–8
B2	$x^2 + y^2 \equiv 1 \pmod 4$ if x, y have no common factor	Stated p. 112 with proof (see also p. 130)
B3	Any cube $\equiv 1$, 0, or $-1 \pmod 9$	Stated p. 128 with proof
B4	Wilson's Theorem: $(p-1)! \equiv -1 \pmod p$ if p is prime	Stated p. 153 Proof (Euler–Gauss) pp. 154–5 Proof (Lagrange) pp. 163–8
B5	Fermat's Theorem: $x^p \equiv x \pmod p$ if p is prime	Stated p. 129 Proofs for $p = 3, 5$ pp. 128–9 general (Euler, "chain") pp. 156–7 general (Euler, algebraic) pp. 161–2
B6	Extended Fermat's Theorem: $g(x)^p \equiv g(x^p) \pmod p$ if p is prime	Stated pp. 211–2 with proof
B7	$f_q^2 \equiv \pm q \bmod (1 + x + \cdots + x^{q-1})$ Where $q \equiv \pm 1 \pmod 4$ and f_q is a certain polynomial	Stated p. 210 (definition of f_q p. 205) Proofs: for $q = 3, 5$ pp. 203–4 for $q = 7$ left to reader (p. 204) general (Gauss) pp. 204–7
B8	Factors of $2^n - 1$ and $2^{2a} + 1$	See G1, G2

LABEL	THEOREM	REFERENCE
C1	Two congruences added	Stated p. 119 with proof
C2	Two congruences subtracted	Stated pp. 119–120 with proof
C3	A congruence multiplied by an equality	Stated p. 118 with proof
C4	Two congruences multiplied	Stated p. 120 with proof
C5	A congruence divided by an equality	Stated p. 122 with proof
C6	Lagrange's Theorem: number of roots of a polynomial	Stated p. 167, proof omitted
D1	Impossible that $2a^2 = b^2$	Stated p. 44 with proof (ancient) Alternate proof (based on F2) pp. 96–7 Alternate proof (based on H3) p. 101
D2	Impossible that $x^2 + 2 = u^3$, unless $u = 3$	Stated p. 148 Proof (Euler) pp. 147–8 (one step omitted)
D3	Impossible that the area of a right triangle be a square number	Stated pp. 104–7 with proof (Fermat)
D4	Fermat's Last Theorem: Impossible that $x^n + y^n = z^n$ if $n > 2$	Stated pp. 107–9 Proofs: for $n = 3$ (Euler) pp. 148–52 (one step omitted) for $n = 4$ (Fermat–Euler) pp. 104–7 general claimed in 1993
D5	Impossible that there be a highest prime	See E4
E1	$2^{13} - 1$ is prime	Stated pp. 132–3 with proof
E2	$2^{16} + 1$ is prime	Stated pp. 133–4 with proof
E3	$2^{32} + 1$ is not prime (factor 641)	Stated pp. 134–5 with proof
E4	There are infinitely many primes	Stated p. 45 with proof (Euclid)
E5	Wilson's criterion for primes	See B4

LABEL	THEOREM	REFERENCE
F1	Greatest common factor of two numbers (GCF theorem)	Stated p. 61 with proof
F2	Unique factorization of any number (Fundamental theorem of arithmetic)	Stated p. 55, proof pp. 60–4
F3	Sum of factors of any number (formula)	Stated p. 70, proof pp. 68–70
F4	Any even perfect number $= 2^a(2^{a+1} - 1)$ where second factor is prime	Stated pp. 73–4 Proof (Euler) pp. 71–3
F5	If $mn =$ square and m, n have no common factor, then m, n are squares	Stated p. 97, proof pp. 96–7
F6	If $mn =$ cube and m, n have no common factor, then m, n are cubes	Stated by implication p. 150 Proof left to reader (see F5)
G1	Any factor of $2^n - 1$ (where n is prime) must $\equiv 1 \pmod{n}$	Stated p. 131 with proof (Euler)
G2	Any factor of $2^n + 1$ (where $n = 2^a$) must $\equiv 1 \pmod{2n}$	Stated p. 133 with proof (Euler)
G3	Any factor of $x^2 + y^2$ (where x, y have no common factor) must *not* $\equiv 3 \pmod{4}$	Stated p. 113, again p. 130 Proof (Euler, by Fermat's Theorem) p. 130 Proof (induction) pp. 136–8, second version pp. 138–9
G4	Any factor of $x^2 + y^2$ (as in G3) must $= u^2 + v^2$	Stated p. 139 Proof (Euler) pp. 140–2 and see last proof of G3
G5	Any factor of $x^2 + 2y^2$ (as in G3) must $= u^2 + 2v^2$	Stated p. 144 Proof (Euler) pp. 142–4 and see last proof of G3
G6	Any *odd* factor of $x^2 + 3y^2$ (as in G3) must $= u^2 + 3v^2$	Stated p. 147 Proof (Euler) pp. 146–7 and see last proof of G3
G7	Any factor of $ax^2 + bxy + cy^2$ (as in G3) must $= Au^2 + Buv + Cv^2$ where $B^2 - 4AC = b^2 - 4ac$	Stated p. 183 Proof (Lagrange) not given
G8	False theorem on factors of $x^2 + ny^2$	See L2

LABEL	THEOREM	REFERENCE
G9	Factors of an even perfect number	See F4
G10	$2^{12}(2^{13} - 1)$ is a perfect number	Corollary of F4 and E1
H1	$m^2 - n^2$, $2mn$, $m^2 + n^2$ make a right triangle	Stated p. 95 with proof (Euclid)
H2	Any primitive right triangle obeys H1	Stated p. 98, proof pp. 97–8
H3	To get a, b so that $a^2 - 2b^2 = \pm 1$, start from $a = 1$, $b = 0$ and let $a' = a + 2b$, $b' = a + b$ etc.	Stated p. 100 with proof (Plato) Proof that no other solutions exist p. 103
H4	Multiplication theorem for $x^2 + ny^2$: The product of two such numbers is also $u^2 + nv^2$	Stated with proof: for $n = 1$ (Diophantus), p. 112 for $n = 2$, p. 142 general (Euler), p. 144 Also a theorem for $ax^2 + bxy + cy^2$ accompanies G7
H5	Binomial Theorem: $(x + y)^n = x^n + nx^{n-1}y + n(n-1)x^{n-2}y^2/2! + \cdots + y^n$	Stated pp. 161–2 Proof (Newton) not given
H6	Formula for sum of factors of a number	See F3
H7	Formula for perfect number	See F4
	In category I, p and q stand for prime numbers, (q/p) for the Legendre symbol	
I1	$(-1/p) = -1$ if $p \equiv -1 \pmod 4$	Follows from G3
I2	$(-1/p) = 1$ if $p \equiv 1 \pmod 4$	Stated p. 140, differently p. 182 Proof (by Lagrange's Theorem) pp. 171–2 Euler's original proof (similar) not given Proof (Lagrange, by Wilson's theorem) p. 172 Also follows from I4 or J1

LABEL	THEOREM	REFERENCE
I3	$(\pm q/p) = 1$ if $p \equiv 1 \pmod q$, where $q \equiv \pm 1 \pmod 4$	Stated for $q = 3$, p. 203 with proof (Euler) for $q = 5$, pp. 203–4 with proof (Euler–Lagrange) for $q = 7$, p. 204, proof sketched (Euler–Lagrange) general p. 204, proof (Gauss) pp. 204–7
I4	Legendre's criterion: $(q/p) = q^{(p-1)/2}$	Stated p. 187 Proof (Euler–Legendre) pp. 186–7
I5	Gauss' Lemma: $(q/p) = (-1)^k qp$	Stated pp. 199–200 Proof (Gauss) pp. 198–9
I6	If $(-pq/r) = (-qr/p) = (-rp/q) = 1$ then $px^2 + qy^2 + rz^2 = 0$ for some x, y, z	Stated p. 193 Proof (Legendre) not given
I7	Quadratic reciprocity: $(p/q) = (q/p)$ except when $p \equiv q \equiv 3 \pmod 4$	Stated p. 191 Proof (Legendre, incomplete) sketched pp. 192–4 Proof (Gauss, infinite descent) sketched pp. 194–6 Proof (Gauss, quadratic forms) not given (see p. 197) Proof (Gauss, by Gauss' Lemma) pp. 198–202 Proof (Gauss, by $f_q(x)$) pp. 203–14
J1	Primitive root theorem	Stated p. 176 Proof (Euler–Gauss) pp. 176–8
J2	Fermat's Theorem and extended Fermat's Theorem	See B5, B6

LABEL	THEOREM	REFERENCE
J3	Lagrange's Theorem	See C6
J4	Legendre's Criterion	See I4
K1	Every "upper" prime is a sum $x^2 + y^2$	Stated p. 113 Follows from G4 and I2 (see pp. 139–40)
K2	Other theorems on $x^2 + y^2$	See B2, G3, G4, H4, I1, I2
K3	Theorems on $x^2 + 2y^2$	See G5, H4
K4	Theorems on $x^2 + 3y^2$	See G6
K5	Theorems on $x^2 + ny^2$	See H4, L2
K6	Theorems on $x^2 - 2y^2$	See D1, H3
K7	Theorems on $ax^2 + bxy + cy^2$	See G7, H4
L1	Goldbach's postulate: every even number = prime + prime	Stated pp. 36–7 Proof not known
L2	False generalization of G4–6 to $x^2 + ny^2$	Stated p. 145 False proof indicated pp. 144–5 Flaw exposed p. 145
L3	All people have same birthday	Stated p. 41 with false proof Flaw exposed p. 42
L4	Fermat's Last Theorem	See D4
L5	Legendre's incomplete proof of quadratic reciprocity	See I7
L6	$41 + 2 + 4 + 6 + \cdots$ generates prime numbers only	Conjectured and refuted, p. 35

Index of Mathematicians

Appendix I

p. 28

I have left this embarrassing admission as I wrote it for the original edition. Shortly after it went to press I realized (as two readers subsequently reminded me) that I did know such a method. Let $x = \frac{1}{2}(1 + \sqrt{5})$, $y = \frac{1}{2}(1 - \sqrt{5})$; then the n'th number of (c_3) is $(x^{n+1} - y^{n+1})/\sqrt{5}$. You don't need the numerical value of $\sqrt{5}$, as it cancels out in the algebra. Thus for $n = 2$ we have

$$
\begin{aligned}
x^3 &= \tfrac{1}{8}[(1 + 3\sqrt{5} + 15 + 5\sqrt{5}) \\
y^3 &= \tfrac{1}{8}(1 - 3\sqrt{5} + 15 - 5\sqrt{5}) \\
(x^3 - y^3)/\sqrt{5} &= \tfrac{1}{8}(8\sqrt{5} - (8\sqrt{5}))/\sqrt{5} = 2,
\end{aligned}
$$

which indeed is the second number. You can prove that it always works by using mathematical induction (described later in this chapter) and verifying that $x^2 = x + 1$, $y^2 = y + 1$.

p. 58

These rules are violated in radioactivity and in reactions due to the bombardment of matter by energetic particles. Such reactions did not play a part in the alchemists' experiments.

p. 59

Actually there are two substances called gold chloride. The more common $AuCl_3$ is red, but the other AuCl is yellow.

p. 109

A proof was announced in May 1993, the year of this revised edition. At the time of writing it appears that the profession has not reached a final verdict on its correctness.

p. 121

Since we are using a twelve-hour modulus, it may alternatively be an hour after midnight.

p. 153

The theorem was first mentioned in a book that attributed it to "John Wilson," but no one seems to know who he was.

Appendix II: Euler's Missing Lemma

This lemma can be stated most intelligibly if we introduce complex integers in the way Euler did, pointing out that

$$a^2 + 2b^2 = (a + b\sqrt{-2}\ )(a - b\sqrt{-2}\ ).$$

Then the Multiplication Rule

$$a = a_1a_2 - 2b_1b_2$$
$$b = a_1b_2 + b_1a_2$$

is secretly a way of saying that

$$a + b\sqrt{-2} = (a_1 + b_1\sqrt{-2}\ )(a_2 + b_2\sqrt{-2}\ )$$

as you can verify by multiplying out the right-hand side. Of course it follows that

$$a^2 + 2b^2 = (a_1^2 + 2b_1^2)(a_2^2 + 2b_2^2)$$

as was shown on page 142 with different names for the variables.

In this book, however, I have not talked about complex integers. So I'll achieve the same thing by saying that if $a^2 + 2b^2 = N$ then the pair (a, b) is a *representation* of the number N; and if $a^2 + 2b^2 = a'^2 + 2b'^2$, but without having $a = a'$ and $b = b'$, then (a, b) and (a', b') are two *different* representations of the same number. And then we can say that any two pairs (a_1, b_1) and (a_2, b_2) can be "multiplied" yielding

$$(a_1, b_1) \cdot (a_2, b_2) = (a, b)$$

where this equation stands for the Multiplication Rule given above. (Note that this relation *implies* $(a_1^2 + 2b_1^2)(a_2^2 + 2b_2^2) = a^2 + 2b^2$ but does not *follow* from it.)

The "Missing Lemma" for the first theorem mentioned on page 147 says that if a and b have no common factor, and (a, b)

represents a cube, so that $a^2 + 2b^2 = r^3$, then we can find m, n so that $(a, b) = (m, n)\cdot(m, n)\cdot(m, n)$. If you apply the Multiplication Rule twice, you find that this means

$$a = (m^2 - 2n^2)m - 2(mn + nm)n = m^3 - 6mn^2$$

$$b = (m^2 - 2n^2)n + (mn + nm)m = 3m^2n - 2n^3.$$

Note that this says *more* than that $r = m^2 + 2n^2$.

(In terms of complex integers, the Missing Lemma says that if a and b have no common factor, and $a^2 + 2b^2$ is a cube, then $a + b\sqrt{-2}$ is a cube. But you can ignore that interpretation if you wish.)

For the second theorem mentioned on page 147, we shall need a corresponding lemma in which $\sqrt{-2}$ is replaced by $\sqrt{-3}$ in the definitions. Thus (a, b) "represents" $a^2 + 3b^2$, not $a^2 + 2b^2$, and the equation $(a_1, b_1)\cdot(a_2, b_2) = (a, b)$ now means

$$a = a_1a_2 - 3b_1b_2$$

$$b = a_1b_2 + b_1a_2$$

Thus the product $(a_1, b_1)\cdot(a_2, b_2)$ has a different meaning from before. The new "Missing Lemma" has the same form as the old, now saying that if $a^2 + 3b^2 = r^3$, and a, b have no common factor, then we can find m, n so that $(a, b) = (m, n)\cdot(m, n)\cdot(m, n)$, but now this means (as follows from the new Multiplication Rule)

$$a = (m^2 - 3n^2)m - 3(mn + nm)n = m^3 - 9mn^2$$

$$b = (m^2 - 3n^2)n + (mn + nm)m = 3m^2n - 3n^3.$$

We now give a proof of the Missing Lemma that applies to both cases, provided that in the second case $a^2 + 3b^2$ is odd; that is all we shall need for the theorem about $x^3 + y^3$.

We shall use the unique factorization theorem proved (for ordinary integers!) in Chapter 3, and the factor theorem proved in Chapter 7, that if p is a prime factor of some $a^2 + 2b^2$ then p itself can be represented as $m^2 + 2n^2$ (likewise for odd prime factors of $a^2 + 3b^2$).

We shall also use a "division theorem" that says that if p is prime, and p is represented by (m, n) and pq by (a, b) then q has a representation (u, v) such that (a, b) is equal either to

$(m, n) \cdot (u, v)$ or to $(m, -n) \cdot (u, v)$. The way to prove it is to note that

$$m^2b^2 \equiv -An^2b^2 \equiv n^2a^2 \qquad \text{(mod p)}$$

where $A = 2$ or 3 depending on which case we are doing; and therefore $m^2b^2 - n^2a^2$ is divisible by p. (This doesn't work if $p = A$, but then we must have $m = 0$, $n = 1$, and u, v are found easily.) So either $mb + na$ or $mb - na$ is divisible by p. Suppose the first. Then if we put $(U, V) = (a, b) \cdot (m, n)$ we find that V is divisible by p, and so is U since $U^2 + AV^2 = p \cdot pq = p^2q$. So we have $U = pu$, $V = pv$, and $u^2 + Av^2 = q$. Furthermore

$$\begin{aligned}(U, V) \cdot (m, -n) &= [(a, b) \cdot (m, n)] \cdot (m, -n)\\ &= (a, b) \cdot [(m, n) \cdot (m, -n)]\\ &= (a, b) \cdot (p, 0)\\ &= (pa, pb)\end{aligned}$$

and therefore $(u, v) \cdot (m, -n) = (a, b)$ as desired. If instead $mb - na$ is divisible by p, then in the same way we let $(U, V) = (a, b) \cdot (m, -n)$ and find $(u, v) \cdot (m, n) = (a, b)$.

In the work above, you can easily check that $(m, n) \cdot (m, -n) = (m^2 + An^2, 0) = (p, 0)$ and also that $(a, b) \cdot (p, 0) = (pa, pb)$. But we have also used the associative property of this pair multiplication. That is,

$$[(a, b) \cdot (c, d)] \cdot (e, f) = (a, b) \cdot [(c, d) \cdot (e, f)]$$

which you can check by applying the Multiplication Rule four times. It also follows immediately from the Multiplication Rule that $(a, b) \cdot (c, d) = (c, d) \cdot (a, b)$, and so in any long product the factors can all be shuffled as we like. (This would be obvious if we used complex integers.)

Now we prove the Missing Lemma. Suppose that a, b have no common factor, and that (a, b) represents r^3. Factor r into prime factors $p_1,\ldots,p_s$ (this list may have repetitions, but it doesn't matter). Then $r^3 = p_1^3 \cdots p_s^3$. By the factor theorem we can find a representation (m_1, n_1) for p_1, and by the division theorem we can get $(a, b) = (m_1, n_1) \cdot (u, v)$, changing the sign of n_1 if necessary. Now (u, v) represents r_3 / p_1 which is still divisible by p_1^2. So we can apply the division theorem twice more, obtaining

$$(a, b) = (m_1, n_1) \cdot (m_1, n_1') \cdot (m_1, n_1'') \cdot (x_1, y_1)$$

where (x_1, y_1) represents $p_2^3 \cdots p_s^3$. In this expression the integers n_1' and n_1'' must be equal either to n_1 or to $-n_1$. But if either of them is $-n_1$, then (a, b) contains the factor $(m_1, n_1) \cdot (m_1, -n_1) = (p_1, 0)$, which makes both a and b divisible by p_1, contrary to our assumption. Therefore $n_1 = n_1' = n_1''$, and we have

$$(a, b) = (m_1, n_1) \cdot (m_1, n_1) \cdot (m_1, n_1) \cdot (x_1, y_1).$$

Now we do the same thing again, extracting p_2 three times from (x_1, y_1) to obtain

$$(x_1, y_1) = (m_2, n_2) \cdot (m_2, n_2) \cdot (m_2, n_2) \cdot (x_2, y_2),$$

where (m_2, n_2) represents p_2.

Doing this for $p_3,\ldots,p_s$ in turn, we obtain

$$(a, b) = (m_1, n_1)^3 \cdot (m_2, n_2)^3 \cdots (m_s, n_s)^3 \cdot (x_s, y_s),$$

where for brevity I have written $(m_1, n_1)^3$ for $(m_1, n_1) \cdot (m_1, n_1) \cdot (m_1, n_1)$, etc.

The pair (x_s, y_s) must represent 1 since all the factors p_j^3 have been extracted. But this is possible only if $y_s = 0$, $x_s = \pm 1$. It is easily seen that $(m_s, n_s)^3 \cdot (\pm 1, 0) = (\pm m_s, n_s)^3$, and therefore by changing the signs of m_s and n_s if necessary we can omit the factor (x_s, y_s).

Rearranging the remaining factors, we now have

$$(a, b) = (m, n) \cdot (m, n) \cdot (m, n)$$

where $(m, n) = (m_1, n_1) \cdot \ldots \cdot (m_s, n_s)$, and the Missing Lemma is proved.

(This lemma would follow right away if we assumed that unique factorization applies to the representations (a, b), or what is the same thing, to the complex integers $a + b\sqrt{-A}$. This is true for $A = 1, 2, 3$ but not for all A; a well-known counterexample is that $(2 + \sqrt{-5})\,(2 - \sqrt{-5}) = 3 \cdot 3$ although none of the factors can be further factored. In Article 169 of the *Elements*, Euler says that $x + y\sqrt{-1}$ must have complex factors if $x^2 + y^2 = (x + y\sqrt{-1})\,(x - y\sqrt{-1})$ has ordinary ones. Later he makes similar statements about $\sqrt{-2}$ and $\sqrt{-3}$. He was presumably appealing to unique factorization for these entities, but he did not prove it. Before blaming Euler for sloppiness, we should remember that his *Elements* is a textbook, not a research report like Gauss' *Disquisitions*. Its aim is to teach you how to solve problems, rather than establish the truth of new discoveries.)

A CATALOG OF SELECTED
DOVER BOOKS
IN ALL FIELDS OF INTEREST

A CATALOG OF SELECTED DOVER BOOKS IN ALL FIELDS OF INTEREST